Reviews of Environmental Contamination and Toxicology

VOLUME 185

Reviews of Environmental Contamination and Toxicology

Continuation of Residue Reviews

Editor
George W. Ware

VOLUME 185

Springer

Springer
New York: 233 Spring Street, New York, NY 10013, USA
Heidelberg: Postfach 10 52 80, 69042 Heidelberg, Germany

ISBN 978-1-4419-2060-7 e-ISBN 978-0-387-30638-4
ISSN 0179-5953

Printed on acid-free paper.

springeronline.com

Foreword

International concern in scientific, industrial, and governmental communities over traces of xenobiotics in foods and in both abiotic and biotic environments has justified the present triumvirate of specialized publications in this field: comprehensive reviews, rapidly published research papers and progress reports, and archival documentations. These three international publications are integrated and scheduled to provide the coherency essential for nonduplicative and current progress in a field as dynamic and complex as environmental contamination and toxicology. This series is reserved exclusively for the diversified literature on "toxic" chemicals in our food, our feeds, our homes, recreational and working surroundings, our domestic animals, our wildlife and ourselves. Tremendous efforts worldwide have been mobilized to evaluate the nature, presence, magnitude, fate, and toxicology of the chemicals loosed upon the earth. Among the sequelae of this broad new emphasis is an undeniable need for an articulated set of authoritative publications, where one can find the latest important world literature produced by these emerging areas of science together with documentation of pertinent ancillary legislation.

Research directors and legislative or administrative advisers do not have the time to scan the escalating number of technical publications that may contain articles important to current responsibility. Rather, these individuals need the background provided by detailed reviews and the assurance that the latest information is made available to them, all with minimal literature searching. Similarly, the scientist assigned or attracted to a new problem is required to glean all literature pertinent to the task, to publish new developments or important new experimental details quickly, to inform others of findings that might alter their own efforts, and eventually to publish all his/her supporting data and conclusions for archival purposes.

In the fields of environmental contamination and toxicology, the sum of these concerns and responsibilities is decisively addressed by the uniform, encompassing, and timely publication format of the Springer-Verlag (Heidelberg and New York) triumvirate:

Reviews of Environmental Contamination and Toxicology [Vol. 1 through 97 (1962–1986) as Residue Reviews] for detailed review articles concerned with any aspects of chemical contaminants, including pesticides, in the total environment with toxicological considerations and consequences.

Bulletin of Environmental Contamination and Toxicology (Vol. 1 in 1966) for rapid publication of short reports of significant advances and discoveries in the fields of air, soil, water, and food contamination and pollution as well as methodology and other disciplines concerned with the introduction, presence, and effects of toxicants in the total environment.

Archives of Environmental Contamination and Toxicology (Vol.1 in 1973) for important complete articles emphasizing and describing original experimental or theoretical research work pertaining to the scientific aspects of chemical contaminants in the environment.

Manuscripts for *Reviews* and the *Archives* are in identical formats and are peer reviewed by scientists in the field for adequacy and value; manuscripts for the *Bulletin* are also reviewed, but are published by photo-offset from camera-ready copy to provide the latest results with minimum delay. The individual editors of these three publications comprise the joint Coordinating Board of Editors with referral within the Board of manuscripts submitted to one publication but deemed by major emphasis or length more suitable for one of the others.

Coordinating Board of Editors

Preface

The role of *Reviews* is to publish detailed scientific review articles on all aspects of environmental contamination and associated toxicological consequences. Such articles facilitate the often-complex task of accessing and interpreting cogent scientific data within the confines of one or more closely related research fields.

In the nearly 50 years since *Reviews of Environmental Contamination and Toxicology* (formerly *Residue Reviews*) was first published, the number, scope and complexity of environmental pollution incidents have grown unabated. During this entire period, the emphasis has been on publishing articles that address the presence and toxicity of environmental contaminants. New research is published each year on a myriad of environmental pollution issues facing peoples worldwide. This fact, and the routine discovery and reporting of new environmental contamination cases, creates an increasingly important function for *Reviews.*

The staggering volume of scientific literature demands remedy by which data can be synthesized and made available to readers in an abridged form. *Reviews* addresses this need and provides detailed reviews worldwide to key scientists and science or policy administrators, whether employed by government, universities or the private sector.

There is a panoply of environmental issues and concerns on which many scientists have focused their research in past years. The scope of this list is quite broad, encompassing environmental events globally that affect marine and terrestrial ecosystems; biotic and abiotic environments; impacts on plants, humans and wildlife; and pollutants, both chemical and radioactive; as well as the ravages of environmental disease in virtually all environmental media (soil, water, air). New or enhanced safety and environmental concerns have emerged in the last decade to be added to incidents covered by the media, studied by scientists, and addressed by governmental and private institutions. Among these are events so striking that they are creating a paradigm shift. Two in particular are at the center of ever-increasing media as well as scientific attention: bioterrorism and global warming. Unfortunately, these very worrisome issues are now superimposed on the already extensive list of ongoing environmental challenges.

The ultimate role of publishing scientific research is to enhance understanding of the environment in ways that allow the public to be better informed. The term "informed public" as used by Thomas Jefferson in the age of enlightenment conveyed the thought of soundness and good judgment. In the modern sense, being "well informed" has the narrower meaning of having access to sufficient information. Because the public still gets most of its information on science and technology from TV news and reports, the role for scientists as interpreters and brokers of scientific information to the public will grow rather than diminish.

Environmentalism is the newest global political force, resulting in the emergence of multi-national consortia to control pollution and the evolution of the environmental ethic. Will the new politics of the 21st century involve a consortium of technologists and environmentalists, or a progressive confrontation? These matters are of genuine concern to governmental agencies and legislative bodies around the world.

For those who make the decisions about how our planet is managed, there is an ongoing need for continual surveillance and intelligent controls, to avoid endangering the environment, public health, and wildlife. Ensuring safety-in-use of the many chemicals involved in our highly industrialized culture is a dynamic challenge, for the old, established materials are continually being displaced by newly developed molecules more acceptable to federal and state regulatory agencies, public health officials, and environmentalists.

Reviews publishes synoptic articles designed to treat the presence, fate, and, if possible, the safety of xenobiotics in any segment of the environment. These reviews can either be general or specific, but properly lie in the domains of analytical chemistry and its methodology, biochemistry, human and animal medicine, legislation, pharmacology, physiology, toxicology and regulation. Certain affairs in food technology concerned specifically with pesticide and other food-additive problems may also be appropriate.

Because manuscripts are published in the order in which they are received in final form, it may seem that some important aspects have been neglected at times. However, these apparent omissions are recognized, and pertinent manuscripts are likely in preparation or planned. The field is so very large and the interests in it are so varied that the Editor and the Editorial Board earnestly solicit authors and suggestions of underrepresented topics to make this international book series yet more useful and worthwhile.

Justification for the preparation of any review for this book series is that it deals with some aspect of the many real problems arising from the presence of foreign chemicals in our surroundings. Thus, manuscripts may encompass case studies from any country. Food additives, including pesticides, or their metabolites that may persist into human food and animal feeds are within this scope. Additionally, chemical contamination in any manner of air, water, soil, or plant or animal life is within these objectives and their purview.

Normally, manuscripts are contributed by invitation. However, nominations for new topics or topics in areas that are rapidly advancing are welcome. Preliminary communication with the Editor is recommended before volunteered review manuscripts are submitted.

Tucson, Arizona G.W.W.

Table of Contents

Rev Environ Contam Toxicol 185:1–33

Persistent Organic Pollutants (POPs) in Eastern and Western South American Countries

Ricardo Barra, Juan Carlos Colombo, Gabriela Eguren, Nadia Gamboa, Wilson F. Jardim, and Gonzalo Mendoza

Contents

Communicated by Lilia Albert.

R. Barra (✉) · G. Mendoza
Aquatic Systems Research Unit, EULA-Chile Environmental Sciences Center, University of Concepción, PO Box 160-C Concepción, Chile.

J.C. Colombo
Environmental Chemistry Laboratory, Faculty of Natural Sciences and Museum, University of La Plata, Argentina.

G. Eguren
Faculty of Sciences University of La Republica, Montevideo, Uruguay.

N. Gamboa
Faculty of Chemistry, Pontifical Catholic University of Peru, Lima, Peru.

W.F. Jardim
Environmental Chemistry Laboratory, Chemistry Institute, State University of Campinas, UNICAMP, Brazil.

I. Introduction

Persistent organic pollutants (POPs) are a global issue, yet the existing research and monitoring efforts are mainly concentrated in the Northern Hemisphere. Hence, a global bias exists in data distribution and obviously constitutes a limitation for a comprehensive global assessment. Recently, UNEP-Chemicals has released a global report, a first effort of data compilation and analysis, which clearly shows that the Southern Hemisphere is underrepresented in terms of information (UNEP 2003a).

The purpose of this review is to provide data on POP distribution in eight countries from eastern and western South America (Argentina, Brazil, Bolivia, Chile, Ecuador, Paraguay, Peru, and Uruguay; thereafter the Region) to support a global assessment (Fig. 1). Human impact is documented by examples of POP pollution in practically every environmental compartment, and local patterns of POP distribution are also discussed. This review presents data summaries collected from a variety of sources, in particular international peer-reviewed journals, governmental reports, and finally contributions from individual researchers of the different countries belonging to the Region.

II. Pollutant Input into the South American Environment

Quantitative information on primary or secondary sources of POPs is very scarce and fragmentary for the Region. Due to their long-term use in the Region, data for pesticides are relatively more abundant. Polychlorinated biphenyl (PCB) information is fragmentary, whereas dioxin and furan source data are practically nonexistent. POP inventories are still in their early stage of development; i.e., PCB inventory work in Argentina, Chile, and Uruguay began in the year 2000 and will start soon in Brazil this current year.

The survey shows that POP levels are abundant in most environmental receptors of interest compared to source evaluation in the South American region. Additionally, very little data on international POP trading are available, although for some countries the amounts of pesticide imported are available but with no possibility of segregating individual POPs.

Considering the lack of reliable quantitative source data for the Region, values provided in this review should be considered cautiously because they are estimates. To improve the quality of these estimates, more detailed information is needed.

A. Pesticides

Practically all the chlorinated pesticides were widely used, produced, or formulated in the Region in the mid-1950s–1960s. There was a declining trend in the 1980s–1990s due to legal restrictions on production and use. Currently, only mirex is legally used in Uruguay.

Aldrin, Dieldrin, and Endrin Cyclodiene organochlorines were first introduced in the Region in the 1950s and experienced rapid growth in use due to their lower

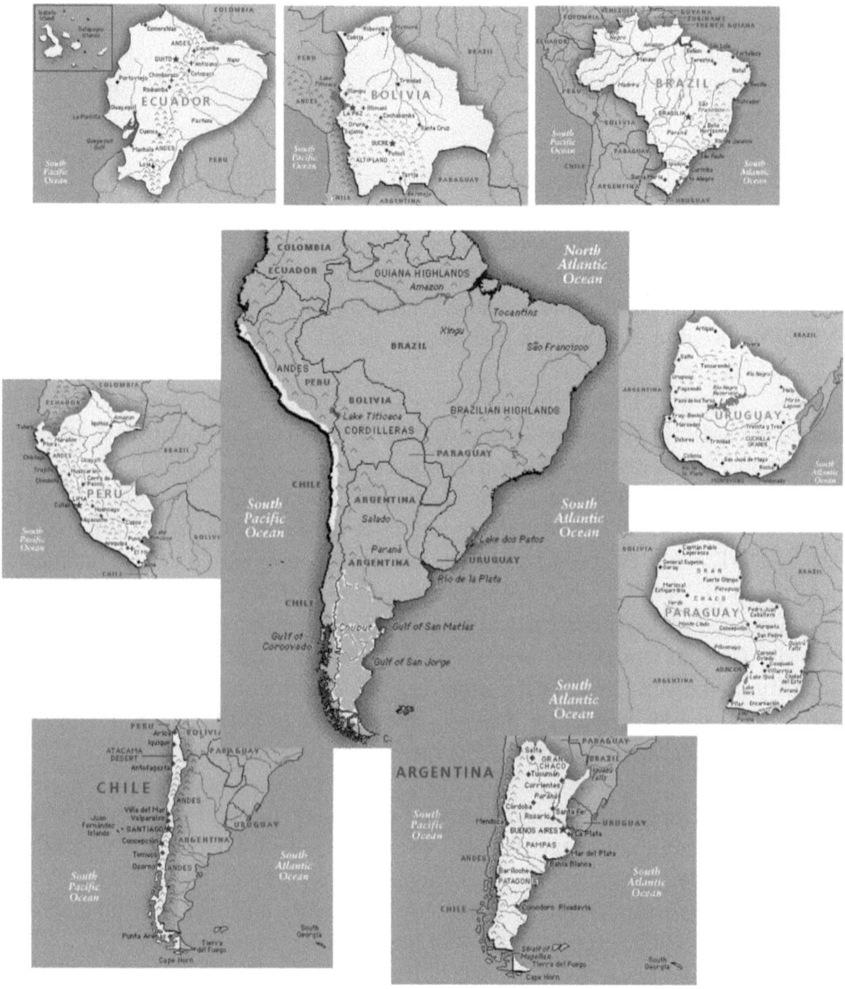

Fig. 1. South American countries considered in this review.

costs and efficacy. They have been formulated and later produced in some countries (i.e., Argentina and Brazil). The Shell Company formulated aldrin and endrin from 1977 to 1990 in the State of São Paulo, Brazil. Despite this production, 294 t aldrin was imported to Brazil from 1989 to 1991 (MDIC 2002). In Argentina, formulation of dieldrin in 1967 was 200 t (5% conc.) and 550 t (18% conc.), compared to 100 t (4% conc.) and 115 t (40% conc.) for aldrin and 300 t (20% conc.) for endrin (Alvarez 1998). In Uruguay, 171 t aldrin, 50 t dieldrin and 60 t endrin were imported from 1968 to 1991 (Boroukhovitch 1999).

Presently, possible sources are stockpiles and severely contaminated areas. For instance, a contaminated site of 600 m^2 in the Shell Company's former site in

Paulinia (Sao Paulo, Brazil) had 1250 t contaminated soil and approximately 750 kg aldrin, dieldrin, and endrin (DRINS). Chlorinated cyclodiene pesticides had widespread use throughout the Region but are now officially forbidden in most countries.

Chlordane Technical chlordane is a very complex mixture of more than 100 chlorinated compounds, with *trans*/*cis*-chlordane and nonachlor as its major constituents. Chlordane was not widely used within the Region. It was formulated and used only in specific cases, especially for ant control. Similar to aldrin, chlordane has been formulated in some countries with Argentina producing 220 t in 1967. Within the Region, no emission measurements have been reported.

DDTs DDT was intensively and widely used in the Region, principally in the early 1960s–1970s for antimalaria programs in tropical and subtropical areas. It was produced as early as 1954 in Argentina (Alvarez 1998). From 1962 to 1982, Brazil produced 73,481 t technical grade DDT. Brazil also imported 31,130 t between 1962 and 1975 and purchased 3200 t between 1989 and 1991 (MDIC 2002). DDT deserves special attention because of its possible reintroduction to the Region to control widespread tropical and subtropical epidemic diseases such as dengue and malaria. Presently, a possible significant secondary DDT source is dicofol. This insecticide is used in the Region and may contain DDT as an impurity. In the year 2000, Brazil imported 111 t dicofol and also produced 209 t. Illegal trade could be another possible source for this POP. No DDT emission measurements have been reported for the Region.

Heptachlor Heptachlor is a nonsystemic insecticide used primarily against soil insects and termites. It degrades into the environment as heptachlor epoxide, which has greater persistency and toxicity than the parent compound. Heptachlor in Brazil has been authorized in agriculture, especially on sugar cane plantations. From 1996 to 2002, around 210 t was imported by Brazil (MDIC 2002). The present estimated stock is around 165 t. There are no emission measurements for the Region.

Hexachlorobenzene The main environmental input of hexachlorobenzene (HCB) occurs as a by-product in the industrial manufacturing of chlorinated solvents as well as some pesticides such as pentachlorophenol. There are well-known, highly contaminated sites and stockpiles of HCB in Brazil, e.g., at Cubatão, which could be a significant environmental source (CETESB 2001). In Cubatâo, HCB was probably generated from perchloroethylene production and from the burning of other chlorinated residues. In 1965, Brazil imported 834 t HCB (MDIC 2002). There are no emission measurements for the Region.

Mirex In Brazil, Uruguay, and Argentina, mirex is popularly known as dodecachlor. It was used principally for ant control with restricted distribution. Mirex has been forbidden in all the Region except Uruguay. There are no emission measurements for the Region. Presently, there is a commercially available product in Brazil called Mirex-S, but the active ingredient is perfluorosulphonamide and not dodecachlor.

Toxaphene Toxaphene is a complex mixture of chlorinated bomanes. There is no record of use for this substance in the Region, and it has been officially forbidden in all countries. There are no emission measurements for the Region.

B. Industrial Chemicals

Polychlorinated Biphenyls PCBs have been intensively used throughout the region, principally in electrical equipment (e.g., transformers). PCB residues from commercial formulations (e.g., Aroclor 1242, 1254, 1260, and equivalent products) are ubiquitous in the environment. Detailed inventory information is still incomplete but there are some country estimates. In Brazil, data are conflicting, indicating 250,000–300,000 t for Askarel (PCB-contaminated oil), which implies a stock of 130,000 t PCBs (Costa 2000). In Chile, PCB stocks have been estimated to be 700 t, 46% of which is still in use (CONAMA 2001). These data are comparable to stocks in Peru (1000 t; González, personal communication, 2002). In Paraguay, 70 t PCB-containing oils are stored in open sites of the National Energy Administration (ANDE). In Uruguay, electrical companies claimed stocks of 81 t PCB-containing oil with 25% still in use (DINAMA 2000). Obsolete electrical equipment and used oil constitute relevant secondary sources of PCBs. An important effort to export PCB-containing material for destruction has been initiated in the Region.

Due to their low water solubility, PCBs have a strong particulate-oriented behavior, adsorb rapidly to suspended matter, and are sequestered in bottom sediments. These residues constitute long-term reservoirs and secondary sources.

Dioxins and Furans PCDDs and PCDFs are ubiquitous trace residues in the environment, resulting principally from combustion processes and chemical production. Air is an important receptor, but the lack of precise measurements in the Region does not permit a detailed analysis. A preliminary regional estimate was calculated considering the correlation between CO_2 emission from fossil fuels and the cement industries (World Resources 2000) and toxicity equivalent (TEQ) emissions to air for some industrialized countries (UNEP 1999), following the approach of Baker and Hites (2000). Figure 2 shows the regression obtained for UNEP countries and estimates from this Region. The regression includes considerable error and does not fully explain the trend of the data ($R^2 = 0.66$). However, the country regression estimates compare well with emission values calculated within the project from activity rates and emission factors using the UNEP (2003b) toolkit. However, more detailed industrial information is considered vital for more precise evaluations (MVOTMA-DINAMA-UNEP 2002). The agreement between these two estimates supports the conclusion drawn.

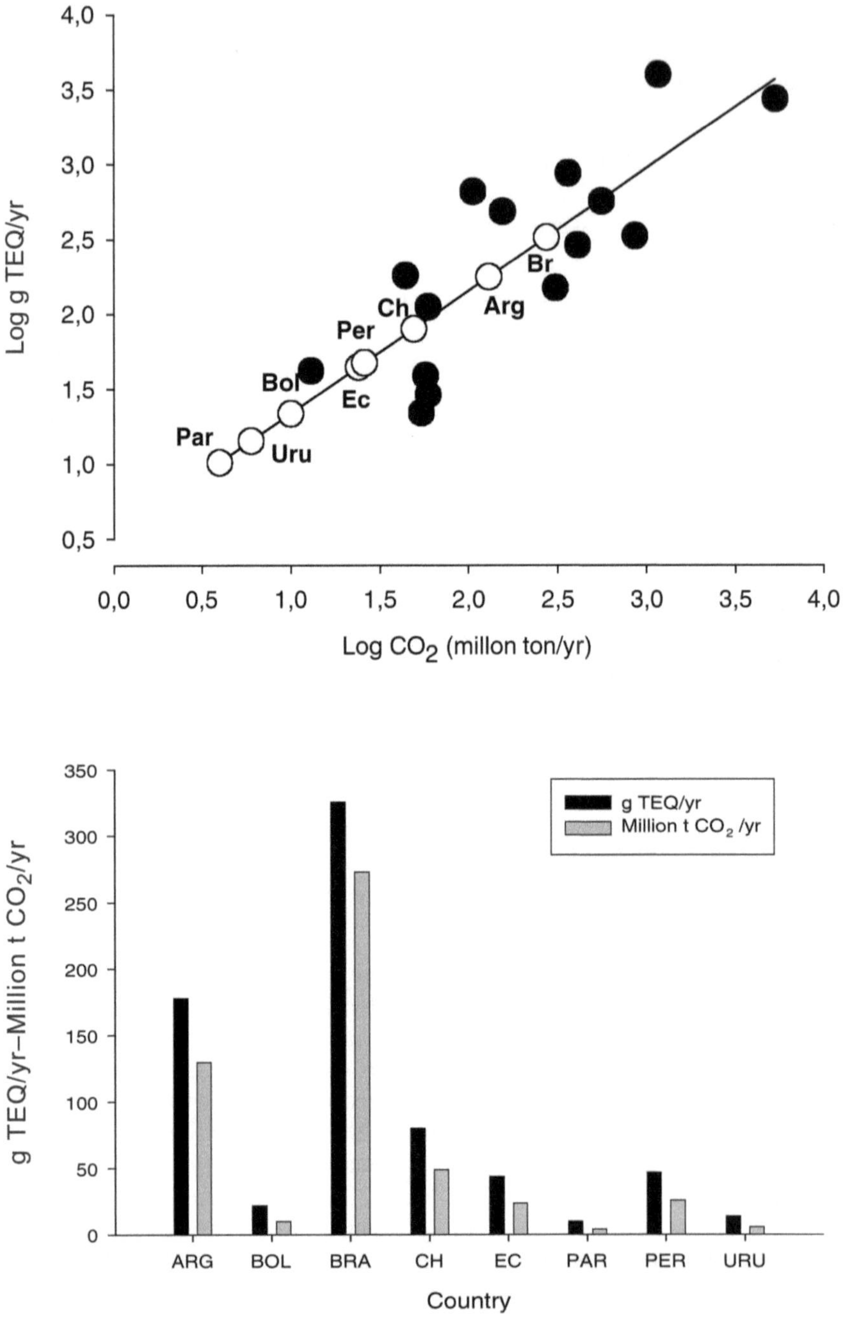

Fig. 2. CO_2–TEQ emission regression for 99 countries (top) and estimated toxicity equivalent (TEQ) emitted to air for the South American countries considered in this study (bottom).

The total regional PCDD and PCDF (PCDD/F) emission to air would be 721 g TEQ/yr, with a markedly uneven distribution because Brazil (45%) and Argentina (25%) together account for 70%. In decreasing order are Chile (11%), Peru (7%), Ecuador (6%), Bolivia (3%), Uruguay (2%), and Paraguay (1%). Individual country emissions are in the intermediate/lowest range of values reported for North America, Asia, and Europe. Considering that the individual country total-to-air ratio from the toolkit estimates are 1.6:2, the total (all media) regional PCDD/F emissions would be of the order of 1300 g TEQ/yr. Toolkit calculations indicate that waste incineration, including hazardous and hospital wastes, is a major source (30%–50%), and comparable to biomass burning (20%–40%). In spite of the agreement of these estimates, validation of emission factors for the Region is considered essential. This estimate should include different industrial technologies, but especially wildfires and biomass combustion, which are relevant processes in South America. Specific combustion experiments with local biomass should be conducted to obtain empirical evidence for local emission factors. At this moment, Brazilian data on PCDD/F emissions are being reevaluated on the basis of recent data on industrial and biomass burning as well as more realistic emission factors.

C. Contaminated Sites as Sources

There are very few officially recognized contaminated sites, and these are located principally in heavily populated industrial areas, i.e., São Paulo, Brazil, Buenos Aires, Argentina, and Santiago and Concepción, Chile. However, these official numbers grossly underestimate the real situation because of the existence of illegal or nonreported contaminated sites throughout the Region. In Argentina, several legal pesticide stocks exist; e.g., the National Health and Food Quality Service (SENASA) has declared stocks of dieldrin (~6 t in La Rioja state) and gamma-HCH (lindane) (1200 L in Cordoba). In the City of Buenos Aires, about 6–30 t HCH and thallium have been stored for years (Vilar de Saráchaga 1997), although recent estimates indicate that this stock would not exceed 10 t. Twenty tons organophosphorous and chlorinated pesticides were inadequately disposed of in El Cuy, Río Negro State, but were remediated in 1998. An obsolete POP stockpile of 6 t residual DDTs from outdated antimalaria programs was deposited in northwest Argentina. The most important known illegal disposal of more than 30 t HCHs and other organochlorines (chiefly DDT) took place in a small town of Argentina in the Province of Santiago del Estero. In Brazil, the environmental agency of the State of São Paulo published a list of more than 600 risk areas (hot spots), the vast majority of which are contaminated by hydrocarbons (CETESB 1997). Three other well-known sites contaminated with miscellaneous chlorinated compounds exist in Paulinia and Cubatão, Sao Paulo, and Cidade dos Meninos, Río de Janeiro (HCH). In Chile, the official disposal of about 100 t PCB-containing oils has been reported in the Antofagasta Region (CONAMA 2001).

III. Environmental Levels

The information is generally aggregated in densely populated areas along the extensive hydrographic basins of major rivers of the Region such as the Amazon, Paraná, and Río de la Plata and consequently the database is strongly biased toward freshwater environments. In spite of the large development of coastal marine areas along the Atlantic and Pacific Oceans, they have received proportionally less attention. Another general regional trend for environmental POP information is the uneven contribution between countries, which reflects the different patterns of economic and technical development.

A. Air

There is no regional routine monitoring program for POPs in air. Available data represent a few geographical areas within the Region, which contrasts with other environmental media and reflects the lesser attention given to air monitoring. The low representation of these measurements should be stressed given the well-recognized high variability of air masses that cross the Region. Table 1 presents POP data available for air.

PCBs Few measurements of PCBs in air have been undertaken in the Region. Recent PCB reports for Santiago de Chile indicated values from 1.04 to 1.75 ng/m^3, comparable to other urban areas of the world (CENMA 2001). PCB levels in air particles smaller than 2.5 μm diameter ($PM_{2.5}$) in Temuco, Chile, ranged from 0.67 to 1.7 ng/m^3, while in Santiago the levels ranged from 1.15 to 2.7 ng/m^3 (Mandalakis and Stephanou 2002). Unexpectedly, observed differences between these two cities were not high, in spite of differences in population density, indicating that the atmosphere in these two Chilean cities is fairly polluted with PCBs even when compared with other urbanized areas of the world.

In the southeast Atlantic Ocean, data presented by Ockenden et al. (2001) reveal several PCB congeners in the Falkland Islands (Islas Malvinas). Levels of PCB congeners 28, 52, 101, 153/132, and 138 ranged from 0.13 to 3.1 pg/m^3, with higher concentrations of the less-chlorinated congeners. Air concentrations over the ocean reached more than 100 pg/m^3 for PCB 28 and only 0.67 pg/m^3 for

Table 1. Persistent organic pollutant (POP) ranges in air.[a]

	PCBs	HCHs	Chlordanes	DDTs	Heptachlors	PCDD/F
Particulate (range)	1–3	1–1.5	1–1.5	1–1.5	1–1.5	3–394
Stack stream deposition (range)[b]	5.9–1,344					13,986–18,000

[a]Data are POP concentrations in ng/m^3; PCDD/F in pg-TEQ/m^3.
[b]Deposition expressed in ng/m^2/d.

PCB 138. In addition, a temporal trend of increasing air levels in summer periods was also reported, with variable concentrations according to the sampling station. Average total PCB levels add up to 5.1 pg/m^3, which illustrates either local sources or the impact of air or water transport to remote environments throughout the Southern Hemisphere.

Dioxins and Furans There are very few measurements of PCDD/F air levels within the Region. Limited data are reported for Brazil, although an interesting study performed by Lohmann et al. (2001) along a transect from Lancaster, U.K, to the southern Atlantic Ocean reported data from coastal and offshore areas in Argentina, Brazil, and Uruguay. In general, reported ambient air concentrations are comparable to those found in remote areas; in fact, in the Falkland Islands (Islas Malvinas), PCDD levels were 330 fg/m^3 for Cl_4–Cl_8 congeners, very similar to remote areas of the Irish coast. The sole exception was Montevideo (Uruguay), which presents a very high, anomalous value of 4000 fg/m^3 of Cl_4–Cl_8 PCDD/F, equivalent to 40 fg TEQ/m^3. Further work is needed to identify the principal sources in this area.

Data reported for Brazil (Krauss 2000) also indicate low PCDD/F concentrations, between 3 and 839 fg *International toxicity equivalents* (I-TEQ)/m^3 while the other countries within the Region (Argentina, Ecuador, Bolivia, Chile, Paraguay, Peru, and Uruguay) do not report data on PCCD/Fs in air.

As a general conclusion, it appears that very low PCDD/F levels are detectable in air samples, but data comparison is difficult because they represent only a little isolated information.

B. Soils

Soils are natural sinks for persistent and lipophilic compounds that strongly adsorb to organic carbon and remain relatively immobile in this reservoir (Mackay 2001). Thus, soil is a typical "long-memory" accumulating matrix. Therefore, POP inputs received in the past will persist with very little clearance and long half-lives. Soils receive environmental pollutant inputs via different pathways, including direct pesticide application, atmospheric deposition, application of sewage sludge or composts, spills, erosion from nearby contaminated areas, illegal deposition, and irrigation with contaminated water. In general, it is difficult to determine when soil contamination occurred, but the level in soil tends to reflect the baseline contamination of a region. An historic evaluation could be performed to determine the predominant input pathway, and occasionally pattern analysis may provide further evidence. On the other hand, urban areas exhibit high POP concentrations produced by industrial activities, whereas in rural areas higher levels of pesticide POPs in soils are expected.

Little information has been obtained at the Regional level, and consequently the database is limited and probably does not reflect the actual situation within the entire Region. Most soil data refer to contaminated urban areas, with few studies covering a large spatial area (Table 2).

Table 2. Pollutant ranges and average values in soils.[a]

	HCHs	DDTs	Heptachlors	Aldrin	Dieldrin	Endrin	HCB	Endosulfan	PCDD/F
RANGE	2–109,290	2–7,868,000	2.79–2,145,000	2.8–24,280	4–11,838	1.7–15,897		20–180	0.001–1.607
Mean	18,270	1,124,040	715,002	7,322	1,983	5,300	6,480	100	0.027
SD	44,591	2,973,807	1,238,414	11,548	4,409	9,177		113	

[a]Data are POP (μg/kg dw); PCDDF (ng–TEQ/kg).

Pesticides Only in a few countries is there information available on chlorinated pesticides in soil, and it does not reflect the actual situation of the Region.

The work of Torres et al. (2002a) described the presence of DDT residues in areas located in the southern Amazon region of Brazil. Considering that the use of DDT in the agriculture has been forbidden since 1985, the presence of this compound in higher levels than its main metabolites suggests that current contamination is derived from the more-recent use of this pesticide or another active principle contaminated with DDT.

A national survey of chlorinated pesticides performed during the late 1980s in agricultural areas from Chile indicated a relatively high detection frequency of these compounds, but the concentrations in surface soils were low (INIA 1990).

PCBs Few studies on PCB levels in soil have been carried out in the Region. The available data are scarce and fragmentary. PCBs have been evaluated in samples from 21 Brazilian municipal solid waste sites with different degrees of maturation. For PCB, the average concentration for the sum of the six congeners (28, 52, 101, 138, 153, and 180) was virtually three times lower when compared to German samples, and all were below the value set by legislation (0.2 mg/kg) (Grossi et al. 1998).

In 18 topsoil samples from Uberlândia (Minas Gerais, Brazil), PCB ranges were 0.05–1.25 μg/kg. These results are comparable to or below background concentrations normally found in temperate soils (Wilcke et al. 1999).

Dioxins and Furans Soils from the Amazon Basin contained 0.02–0.4 ng I-TEQ/kg PCDDs and PCDFs. The low PCDD/PCDF concentrations were explained by the absence and lack of dioxin formation during forest burning or by a lack of humic acids in soils, which prevented the adsorption of dioxins. This behavior also reinforces the hypothesis that dioxins measured during biomass burning may result from volatilization and not *in situ* formation (Krauss et al., 1995).

Braga and Krauss (2000) found higher values of dioxin in soil collected from a supposed remediated area from Cidade dos Meninos (Rio de Janeiro State, Brazil), a well-known POP hot spot resulting from stockpiles left from a pesticide industry operated by the Ministry of Health that is now deactivated.

PCDD/F levels in 21 Brazilian municipal solid waste sites showed that most PCDD/F concentrations are above the levels acceptable by German standards (17 ng I-TEQ/kg), especially for samples from metropolitan areas. However, the results are similar to those found for Germany (Grossi et al. 1998).

According to the information provided recently by Krauss (2000), several locations in Brazil present very high dioxin and furan levels in soils, particularly at a former HCH production plant with levels reaching 13,900 ng I-TEQ/kg.

Hot Spots in the Region It is widely accepted that several POP stockpiles are probably relevant sources for soil contamination, but few data have been gathered to date. Some institutions in Chile and Uruguay and the IBAMA in Brazil have begun a project to identify sites contaminated with PCP and other chemicals. Additional data on contaminated soils should soon be reported as these programs are implemented.

In the study performed by Oliveira and Brilhante (1996), the authors demonstrated that the risk of contamination was restricted to the population close to the contaminated area and the consumption of food and other products produced in Duque de Caxias (Rio de Janeiro State, Brazil). In this region, a lindane plant belonging to the Ministry of Health was operated for many years, and when it was closed in 1955, many tons of pesticides were left unattended in the area. An extremely high level of surface soil contamination (of the order of hundreds of mg/kg) was found in the area up to 100 m distant from the old plant.

The high POP values in soils obtained in Canoas (RS) were associated with a pesticide industry. In Paulínia (São Paulo State), Shell Chemical manufactured pesticides from 1975 to 1993, contaminating the soil with aldrin, endrin, and dieldrin.

Another well-known POP (dioxin and furan) hot spot in Brazil refers to the Solvay Company in Santo André, São Paulo State. In the CETESB (2000) report, it was shown that lime samples had dioxin and furan concentrations in the range of 0.1–31, 138 ng/kg TEQ for samples collected in August and December 1998. In March 1998, high dioxin levels were found in milk produced in the German state of Baden-Wurttemberg, leading to its removal from the market. Tracing the origin of this contamination led to cattle feed that was tainted with high dioxin levels. Six components of the feed were analyzed separately and the citrus pulp pellets from Brazil were identified as the source. A broker, Carbotex Industry and Commerce de Cal Ltda., had marketed the lime produced by Solvay in Santo André since 1986.

Rhodia (Cubatão City, São Paulo State) manufactured chemicals used for wood treatment, such as pentachlorophenol and sodium pentachlorophenate as well as tetrachloroethylene and carbon tetrachloride. The principal chemical waste compounds from the manufacture of these chemicals were HCB, hexachloroethene, and hexachlorobutadiene. In 1984, it was reported that the company had 11 illegal waste dumps around the area.

C. Waters

As indicated previously, the majority of information on POP levels is for the major rivers of the Region. Table 3 presents Regional POP ranges and averages for waters. The database is not completely representative for the entire area because its information comes principally from Argentina and Brazil (~95%) and, to a lesser extent, from Uruguay, Ecuador, and Chile. There is a lack of information from the other countries of the Region.

Pesticides Overall, the high pesticide levels in freshwaters of the Region suggest a complex situation, but the small amount of data requires a cautious interpretation. Usually, only suspected contaminated ecosystems are monitored, and large-scale

Table 3. Pollutant ranges and averages in waters.[a]

	PCBs	HCHs	Chlordanes	DDTs	Heptachlors	Aldrin	Dieldrin	Endrin	HCB	Endosulfan	Mirex
Range	7–39	3–790 (7,300)	0.6–400 (7,500)	0.6–6,510	0.9–1,060	1–3,710	0.8–5,000	2–230	1–14,160	1–1,900	0.3
Mean	17	146 (622)	104 (703)	1,267	206	74 (187)	42 (447)	100	418 (1,384)	343	
SD	14	191 (1,572)	124 (1,978)	1,920	274	79 (647)	50 (1,200)	61	422 (3,280)	529	

[a]Data are POP (ng/L). Bracketed values include critical cases.

regional water monitoring programs have never been undertaken. In addition, very frequent nondetectable levels were not included in the database; e.g., mean values in Table 3 overestimate the regional picture due to the strong influence of contaminated sites.

High variability in the pesticide levels (0.6–14,160 ng/L) reflects distinct ecosystem conditions, from less-impacted environments to severely polluted streams located in densely populated and industrialized areas close to Buenos Aires and Sao Paulo. Heptachlor, HCHs, aldrin, and DDTs are the most frequently reported pesticides in water, accounting for more than 60% of the total database.

As expected, according to its higher water solubility, lindane and its isomers are frequently detected at high levels. The highest mean is 622 ± 1572 ng/L, and the information principally reflects Argentinean data (47 of 56 cases). Indeed, high HCH concentrations (0.3–7 μg/L) in the Reconquista River, Buenos Aires (Topalián et al. 1996; Rovedatti et al. 2001) and in some northern and central areas (García Fernandez et al. 1979; Caviedes Vidal 1998) exceed by 3–400 times the Canadian guidelines (10 ng/L).

Concentrations as high as 520 ng/L have been reported for the Río de la Plata near the Buenos Aires sewer, and data from important Argentinean rivers in the South (Paraná, Uruguay, Río de la Plata) or the north of the country (Río Negro, Río Limay) are consistently higher than guidelines. The relatively high pesticide levels in the Río Negro, Patagonia, reflect the intense agricultural activity in this valley (vegetables and fruits) and related pesticide use, e.g., the organophosphates dicofol, endosulfan, endrin, and heptachlor (Natale et al. 1988). Finally, excluding these critical sites, the general mean of HCH decreases drastically to 146 ± 191 ng/L, although it is still higher than the guidelines, suggesting that these compounds are of concern throughout the country.

Heptachlor and aldrin data are more widely distributed within the region and have lower general means than HCHs (206 ± 274 and 187 ± 647 ng/L, respectively) but are still higher than guideline values (3.8–4 ng/L; Canadian EPA). Dieldrin presents a worse situation, especially in Brazil, which has a larger set of reports (12 of 19 cases) and includes the highest levels. The general dieldrin average (447 ± 1200 ng/L) is almost an order of magnitude higher than the USEPA guideline (56 ng/L). The general means of aldrin and dieldrin are shifted by a few contaminated sites in Brazil (3710 and 1000–5000 ng/L, respectively; CETESB 1997; Chagas et al. 1999) and Argentina (dieldrin, 1429 ng/L; Caviedes Vidal 1998). Excluding these outliers, the averages decrease 3–10 times (42–74 ng/L), closer to the guidelines.

DDTs are also frequently reported at very high concentrations, and the general mean (1267 ± 1920 ng/L) is double that of the more-soluble HCHs. Reports from Argentina and Brazil indicate very high levels (1000–6000 ng/L), more than three orders of magnitude above the USEPA guideline (1 ng/L), which could reflect higher inputs related to subtropical agricultural exploitations and vector control. However, this variability probably includes methodological uncertainties, especially in older works, because water analysis of highly hydrophobic compounds such as DDTs is recognized as more difficult. As observed for HCH, excluding the critical samples, the DDT average decreases to a few tens of ng/L, but still remain 1–2 orders higher than guidelines.

HCB concentrations present the highest regional average, similar to DDTs, but with a larger variation in levels ($1{,}384 \pm 3{,}280$ ng/L), that are within accepted guidelines (USEPA: 3680 ng/L). This average is raised by Brazilian data, which have the largest HCB data set (25 of 31 cases) with the highest concentrations (6,000–14,160 ng/L; Celeste and Cáceres 1987; CETESB 1997).

PCBs PCBs have only been reported in Brazil (Paraiba River, Rio de Janeiro), Chile (Biobío River), Argentina, and Uruguay (Uruguay River and Río de la Plata). Available data are scarce, and the concentrations detected range from non-detectable to higher than the guideline value.

In the Uruguay River, the recorded concentrations (7 ng/L) are above the recommended Argentinean limit (1 ng/L) (DINAMA-SOHMA-SHN 1998), and the PCB levels in the Río de la Plata River (close to Buenos Aires) and the Biobío River (22 ng/L) are even higher than this more-permissive guideline (USEPA recommend a guideline value of 14 ng/L) (Gavilan et al. 2001). In Brazil, Telles (2001) showed that PCB and organochlorine compound concentrations in surface water samples collected in the State of Pernambuco were below the detection limit. Another important survey carried out in Brazil by UFSCar/UNICAMP/CETESB (Qualised Project) showed nondetectable PCB levels in the water column and interstitial waters of the Tiete River.

Dioxins and Furans The recent environmental survey from CETESB (2002) in the Santos Bay (São Paulo State, Brazil) is the only published study for PCDD/Fs in water, and all 25 analyzed samples were below the detection limit.

D. Sediments

Table 4 presents POP ranges and averages for freshwater sediments. In this case, the database ($n = 214$) presents a more-uniform contribution from the different countries: Argentina (45%), Brazil (31%), Chile (14%), Peru (10%), and Uruguay (5%). Individual POP information is also more equilibrated, but chlorinated pesticide data still dominate (61%) followed by Polycyclic aromatic hydrocarbons PAHs (23%), PCBs (10%), and PCDD/F (6%). Two reports for Peruvian coastal sediments (Whelan and Hunt 1983) and 10 from the Brazilian CETESB (2001) for an environmental survey in the Santos Bay (São Paulo State) were included that also include organic mercury in sediments from the Amazon region.

Overall, as observed for water data, sediment data indicate a complex situation in densely populated areas affected by urban-industrial inputs and that present high POP levels. The most frequently reported POPs in sediments are DDTs, HCHs, PCBs, and heptachlors, accounting for 46% of the total database. The concentrations indicate a large variability, basically introduced by some highly contaminated sites in Argentina and Brazil, which present levels 4–5 orders of magnitude higher (Costa 1997; Tavares et al. 1999; Torres et al. 2002b).

Table 4. Pollutant ranges and averages in sediments.[a]

	PCBs	HCHs	Chlordanes	DDTs	Heptachlors	Aldrin	Dieldrin	Endrin	HCB	Endosulfan	Mirex	PCDD/F
Range	0.8–23 (580)	0.1–15 (57,100)	0.2–15	0.1–48 (85,800)	0.1–17 (2,700)	0.02–4	0.2–5 (24,000)	0.2–2 (22,200)	0.1–22 (26,400)	0.5–49 (23,500)	0.03	0.04–5.12
Mean	9.1 (58)	3.2 (4,395)	4.2	9.7 (4,009)	3.9 (211)	1.4	1.0 (1,847)	0.8 (3,701)	6 (4,473)	9.4 (14,008)		
SD	7.7 (149)	4.4 (15,836)	5.7	14 (18,274)	5.4 (748)	1.5	1.3 (6,656)	0.6 (9,063)	13 (10,762)	20 (37,037)		

[a]Data are POP (μg/kg dw); PCDD/F (μg TEQ/g). Bracketed values include critical cases.

Pesticides Essentially, the chlorinated pesticide patterns reveal some very critical sites and more-homogeneous residual concentrations. Thus, the principal mean averages consistently exceed all sediment guidelines. Excluding these more-critical sites (see bracketed values in Table 4), the general averages decrease in 2–3 orders of magnitude. However, they still remain generally higher than Canadian freshwater guidelines for aquatic life protection, i.e., heptachlor (3.9 ± 5.4 vs. a reference value of 0.6 μg/kg); DDTs (9.7 ± 14 vs. 6.15 μg/kg); HCHs (3.2 ± 4.4 vs. 0.94 μg/kg); and chlordanes (4.2 ± 5.7 vs. 4.5 μg/kg).

Recent studies carried out in an urban site in central Chile reported low DDT levels in a dated sediment core. The historic pattern registered a deposition rate of 1309 ng/m^2/yr during 1972–1978, which declined steeply in recent years after DDT banning during the 1980s. However, actual lindane deposition rates are relatively high (188 ng/m^2/yr), reflecting the recent use of this pesticide (Barra et al. 2001). Additionally, Torres et al. (2002b) detected in sediment samples from Paraíba do Sul-Guandu system, average concentration of ΣDDT around 225 ng/g and trace concentrations of HCB.

PCBs Overall, PCB concentrations in sediments are relatively high, but the database is strongly biased by contaminated sites in Argentina, the Río Santiago (998 μg/kg; Colombo et al. 1990), and Brazil (580 μg/kg; Lamparelli et al. 1996). Excluding these sites, the general PCB mean decreases from 58 ± 149 to 9.1 ± 7.7 μg/kg, lower than the Canadian guideline for aquatic life protection (34.1 μg/kg). However, polluted sediments from the Río de la Plata estuary often exceed this value. PCB values in sediments from a remote highland lake in northern Chile have recently been reported (Barra et al. 2004) with levels that are quite low, reaching maximum values of 2 ng/g dw, but with trends indicating an increase in recent times.

E. Biota

As expected, available data on POPs in South America for both aquatic and terrestrial animals are relatively scarce when compared to other regions of the globe. Aquatic organisms are by far the most studied organisms, and among them, bivalves and fish. Terrestrial organism data are almost completely centered on birds, with some Brazilian data available for bovines and insects. As observed for other environmental receptors, regional data distribution is uneven, heavily centered in coastal environments, principally on the Atlantic coast. Data availability reflects both the capacity of each region to maintain a routine environmental monitoring program as well as the environmental monitoring carried out by universities and research centers, which are by far the two most active data generators in South America. The data presented for aquatic and terrestrial organisms are solely from Argentina, Brazil, Chile, Peru, and Uruguay, and most of them are centered in the Amazon basin and estuarine ecosystem (Paraná-Río de la Plata). Moreover, the recent environmental survey in the Santos Bay (São Paulo State) from CETESB reported the first data on PCDD/F concentration in biota within the region (CETESB 2002).

Shellfish Considering the South American coastal environment as a whole, the most comprehensive program of POP monitoring in coastal organisms is the Mussel Watch (Farrington and Tripp 1995). Sampling sites showing high POP levels in the Region include Recife, Brazil (Fig. 3; BRRE), Río de la Plata, Argentina (ARRPi m), and Punta Arenas, Chile (CHPA). Among the POPs reported in this study, PCBs predominate, followed by DDTs and chlordanes, whereas when all data gathering is considered (including >10 yrs old), DDT, HCH, DRINS, and HCB are the most common POPs.

Baseline PCB concentrations range from 200 to 700 μg/kg lipids in unpolluted sites, 1000 to 3,000 μg/kg in samples from moderately contaminated sites, and 4,000 to 13,000 μg/kg lipids in the most affected bivalves. Comparison of the presented values is difficult because the lipid content does not normalize all data. In this context, when considering all data available for bivalves, the PCB measurable concentration averages 20 ng/g in almost all the gathered data, reaching values above 200 ng/g only in the work of Farrington and Tripp (1995). This is an interesting result, and according to Lopes et al. (1992), these high values may reflect a seasonal variation in POP content in bivalves. Fresh weight concentrations in these bivalves are 1 order of magnitude lower than the guideline for PCBs (0.15–0.2 compared to 20 ng/g), and concentration patterns tend to follow

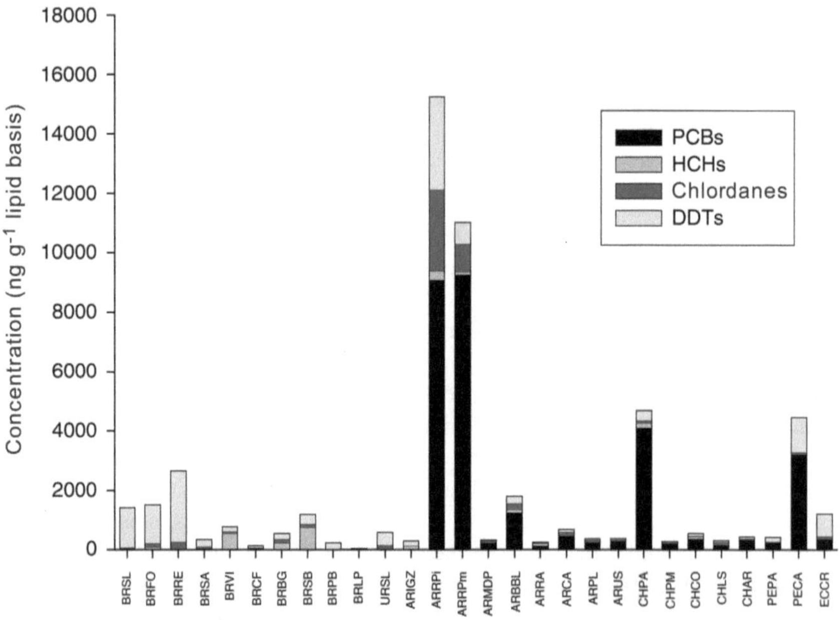

Fig. 3. Mussel Watch data for South America. From the left in north to south direction on the Atlantic (Brazil, BR; Uruguay, UR; Argentina, AR) and from south to north on the Pacific (Chile, CH; Peru, PE; Ecuador, EC).

both the industrial development and population growth observed in the Pacific and Atlantic coastal areas.

Within the framework of the international Mussel Watch Program (Sericano et al. 1995), levels of several congeners of PCBs were analyzed in five samples collected within the wide latitudinal gradient of Chilean coast (see Fig. 3), indicating very low levels, increasing as one moves south. Average DDT values are around 10 times lower than those observed for PCBs, well below the proposed guideline value (5,000 μg/kg), and follow a similar pattern observed for this class of compounds.

Fish The database used to assess the present stage of fish contamination by POPs is composed of a set of reports from Argentina, Brazil, and Chile. The majority of these results are normalized by lipid content, and the trends observed are similar to those obtained in bivalves. POP contents in fish present a tremendous variability resulting from environmental characteristics as well as organ-related components. PCBs are the major organochlorine residue, followed by DDTs and HCHs.

Fish from the Río de la Plata River presented the highest POP levels, principally PCBs and PCDD/F, when compared with samples collected upstream along 1500 km on the Paraná and Iguazú Rivers. In contrast, Focardi et al. (1996) observed PCB levels around 530 ng/g lipid ($n = 45$) in fish muscle sampled upstream in the Biobío River (central Chile), an order of magnitude higher than the values reported for marine fish from the Argentine coast. Additionally, the heavy industrial activity in the Biobio basin (pulp and paper mills, petrochemicals, forestry, and diversified agriculture) is reflected in the high value found in *Mugil cephalus* sampled in the river mouth (1842 $\pm$ 1005; ng g^{-1} lipid basis, $n = 6$).

With respect to chlorinated compounds, the majority of the data refer to the Tietê Basin (São Paulo State).

A very comprehensive survey in the Santos Cubatão area (Brazil) was recently carried out by the São Paulo State Agency (CETESB), selecting 26 sampling points, including water ($n = 26$), sediments ($n = 71$), and biota ($n = 2242$). The results indicate that PCBs are by far the most abundant persistent toxic compound, being detected in 96% of the organisms, with 14% above the human criteria for daily consumption. PAHs rank as the second most important contaminant, being detected in 76% of all organisms, although none was above the acceptable levels for consumption. As expected, the most abundant dioxin found in 11 biota samples was octachlorine (OCDD), detected in 10 samples. TEQ values in the area varied from 0.6 to 1.7 pg/g for crabs, 0.003 for bivalves (*Perna perna*), and 3.45 pg/g for *Cassostrea* sp., and from <0.0001 to 0.16 pg/g for fish (*Mugil curema*). These values are in general low when compared to similar organisms analyzed in Argentina (clams) and in the Northern Hemisphere.

Other Aquatic Organisms Other organisms that have been analyzed for POPs in the Region are crabs (*Cyrtograpsus angulatus*), dolphins (*Pontoporia blainvillei*), and porpoises (*Phocoena spinipinnis*) from the Atlantic Argentine coast and continental shelf (Menone et al. 2001; Borrel et al. 1994; Corcuera et al. 2002). As expected, according to their different feeding habits and trophic status, dolphins and porpoises collected along the Buenos Aires Atlantic coast present ~1 order of magnitude higher lipid concentrations in blubber samples. PCBs show the highest levels (averages, 296, 1980, and 3300 μg/kg lipid in crabs, dolphins, and porpoises, respectively), closely followed by DDTs (170, 1670, and 4320 μg/kg lipid). Cetacean values are comparable to those of moderately contaminated bivalves from the Mussel Watch (Fig. 3) but much lower than those of polluted clams and fish from the de la Plata River.

Samples from *Callinectes sapidus* (an omnivorous crab) collected in the Santos Bay area (Brazil) showed octachlorine dioxin to be the most important congener in these animals (~ 30 pg/g ww), followed by the heptachlor congener (~8 pg/g), whereas for furans, the 7-chlorine congener was the most abundant. Total TEQ values averaged 1.5 pg/g ww.

In Chile, in crustacean and gastropods from the Valdivia area (southern Chile), DDTs and DRINs are the POPs with the highest concentrations. The levels observed are low compared to other regions of the world, ranging from 0.8 up to 5 ng/g ww (Palma-Fleming et al. 1998).

Birds Bird eggs from common gulls were analyzed to correlate PCB levels to different trophic status and anthropogenic influence for different colonies sampled along the Chilean coast. PCB levels were higher by a factor of 2 in central Chile when compared to eggs collected in southern Chile, reflecting the impact of anthropogenic sources. Nevertheless, average data for six PCB congeners indicate that levels are at least 1 order of magnitude lower than those reported for gulls' eggs in the Northern Hemisphere, ranging between 53 and 245 ng/g ww, respectively (Muñoz and Becker 1999).

Focardi et al. (1996) analyzed PCB residues and chlorinated pesticides in muscle and liver from three bird species collected in the Biobío River Basin (central Chile) in an attempt to evaluate a possible pollution gradient from the Andes Mountains (Santa Bárbara) to the Pacific Ocean (380 km). PCB levels were higher in the most urbanized area (Concepción) with important industrial activity. A similar compositional pattern observed for PCB congeners in both areas indicates a common POP source.

F. Foods

Owing to their persistent and bioaccumulative properties, POPs are transferred through terrestrial and aquatic food chains and ultimately accumulate in top predators and humans. Omnivorous humans occupy a top position in terrestrial and aquatic food chains, thus favoring the biomagnification of POPs.

The data collected clearly reflect the regional need to control POP levels in food, especially milk products, grains, vegetables, and oils (e.g., soybean, cotton, corn, fish, sperm, and peanut). At the same time, results reported in older papers should be considered cautiously because of the changes in analytical techniques in the past two decades.

Most frequently reported POPs are DDTs and HCBs (five countries), HCHs (mainly lindane and aldrin (four countries), followed by heptachlors, dieldrin, and endrin (three countries), chlordanes and endosulfan (two countries), and PCBs, mirex, and toxaphene (one country). In contrast, PCDD/Fs and PCBs are scarcely reported for food items in the region. Chile and Argentina are the countries with more reports, whereas no data are available from Paraguay, Bolivia, and Ecuador. Overall, HCHs and DDTs usually present the highest concentrations in food items in the Region, with a decreasing trend in recent years, probably associated with restrictions in the use of these pesticides for agricultural purposes (Table 5).

G. Humans

POPs pose a serious risk to human health because these substances may be accumulated in the adipose tissues leading to body burdens and because they have the potential to be biomagnified through the food chain. Once ingested, POPs are sequestered in body lipids, where they equilibrate at roughly similar levels on a fat-weight basis between adipose tissue, serum, and breast milk. Since the early 1980s, concerns within the Region have increased with respect to POP levels in different human tissues. Several governmental reports and occupational exposure surveys, and to a lesser extent scientific papers, were produced during 1980–1990.

The Region presents important differences with respect to the information obtained both qualitatively and quantitatively. Argentina, Brazil, Chile, and Uruguay have a long history of human tissue sampling with standardized protocols for analyzing most POPs, whereas in Peru fewer analyses of human samples were reported.

Chlorinated Pesticides Chlorinated pesticides were more often reported (76% of the total data) and discriminated by individual pesticides, where the most frequently detected were DDTs, HCHs, HCBs and DRINS (aldrin, dieldrin and endrin). On the other hand, the most often monitored matrices were breast milk and blood, with scarce reports for adipose tissues and urine.

Some chlorinated pesticide data in humans show high levels, with some critical cases for DDT in breast milk and blood and for dieldrin in blood. Breast milk was the matrix that presented the highest variability for all studied compounds, followed by blood (Table 6). This factor, together with differences in statistical management, makes comparison difficult and only permits a discussion based upon range and mean values.

In general, it is evident that the countries of the Region differ in terms of database uniformity for the different POPs in humans. Three groups may be discerned: (a) countries with a broader database (more compounds), e.g., Argentina and Uruguay; (b) countries that focused on a few compounds, e.g., Chile, Brazil, and Ecuador; and (c) a third group with no information on human POP level, e.g., Bolivia, Paraguay, and Peru.

Table 5. Pollutants (μg/kg) in food for human consumption in the South American region.

Food	HCH	Chlordanes	DDT	Heptachlors	Aldrin	Dieldrin	Endrin	HCB	Endosulfan	Mirex	References
Milk[a]	51–360	23	200–990	55		40		7	17		Maitre et al. 1994; Higa 1978;
	56–381		128–486	60–375		30–276	227				Astolfi and Higa 1978
											Ministry of Health (Chile) 1994
Cheese[a]	190		110			50					Higa 1978; Astolfi and Higa 1978
	65–302		9–629	4–212		79–785					Ministry of Health (Chile) 1994
Butter[a]	29–160		24–140	64	110	30					Lenardón et al. 1994; Higa
	31–165		134–721	321–		24–347		6			1978; Astolfi and Higa 1978
	25–85		89–202	906	72–						Ministry of Health (Chile) 1994
				204	347						INDA 1987
Dairy[a]	12	1.7	11.6	29	1.02	1.26	0.43	0.85	0.8		Villaamil 2000
Infant milk	0.65[c]	0.12[c]	0.29[c]–52[b]	0.64[c]	0.4[c]	2[b]		0.14[c]	0.16[c]		Astolfi and Landoni 1978; Rodriguez Girault et al. 2001

Table 5. (Continued)

Food	HCH	Chlordanes	DDT	Heptachlors	Aldrin	Dieldrin	Endrin	HCB	Endosulfan	Mirex	References
Beef	1.7[a]–6[b]	0.15[a]	2.28[a]	16.61[a]	0.59[a]	0.6[a]–		1.03[a]	0.61[a]	2.65[a]	Villaamil 2000; Higa 1978;
	280[b]	149[a]	423[a]	238[a]	107[a]	49[b]					Astolfi and Higa 1978
	136[a]					263[a]					Ministry of Health (Chile) 1994
Pork[a]	0.59	1.62	3.04					0.33			Villaamil 2000
Chicken[a]	5.84	0.11	5.67	2.82	1.19	0.69	0.08		0.67		Villaamil 2000
Eggs	4	5									Ministry of Health (Chile) 1994
Fish	1.63	1.23	8.18	16.85		1.04	7.7	3.5	0.83		Villaamil 2000
Fish oil	3.47		42.3					3			Ballschmitter and Zell 1980
Corn[c]	10										Higa 1978; Astolfi and Higa 1978
Corn oil	43.7		121.3	169	152.2						INDA 1987
								430			Lara et al. 1999
Soya oil								2045			Lara et al. 1999
Wheat[c]	610										Higa 1978; Astolfi and Higa 1978
Cereals[c]	0.05		5.2			0.27	2.21		0.02	0.09	Villaamil et al. 1999
			250					38			Lara et al. 1999a, b
Fruits[b]	0.26	0.1	0.08	0.28	0.03	0.03					Villaamil 2000

Table 5. (Continued)

Food	HCH	Chlordanes	DDT	Heptachlors	Aldrin	Dieldrin	Endrin	HCB	Endosulfan	Mirex	References
			10								Higa 1978; Astolfi and Higa 1978
			30–740								Higa 1978; Astolfi and Higa 1978
Tomato[b]			26								Lara et al 1999.
			320					8	280		Lara et al. 1999
					201						Lara et al. 1999
Carrot					1319			9			Cabrera 1993
Potato			1								Gonzalez et al. 2001;
					279						Villaamil et al. 1999a, b
Vegetables[b]	8–10	0.5–43	0.39–65 4–20	1.33–18	2– 226	0.13	0.07– 218	0.09 12–22	1.4–2	0.2	Lara et al. 1999

[a]Lipids
[b]Fresh weight
[c]Dry weight

Table 6. Pollutant ranges and averages in humans (μg/kg or μg/L).

Sample	DDT	HCHs	HCB	Dieldrin
Breast milk	$n = 76$	$n = 35$	$n = 34$	$n = 31$
Range	9.0–230 (9×10^3)	5.0–150 (3,910)	8–205	3–48
Mean	69.6 (1,109)	86.5 (620.3)	77	30.3
Blood	$n = 48$	$n = 41$	$n = 39$	$n = 29$
Range	0.4–97 (17×10^4)	1.1–50 (308)	3–42	0.7–20.4 (3×10^5)
Mean	25.0 (12,135)	15.0 (39.4)	24.2	6.94 (25,042.9)
Fatty tissue	$n = 5$	$n = 4$	$n = 4$	$n = 3$
Range		0.3–19 (790)	0.4–13 (146)	0.04–0.04 (70)
Mean		6.5 (202.4)	4.6 (39.9)	0.04 (23.36)
Urine	$n = 4$	$n = 2$		
Range	11.3–34.9	14.4–46.3		
Mean	21.9	30.4		

Bracketed values include critical cases.

The exposure pathways to chlorinated pesticides are very similar throughout the Region, and in general terms occupational and accidental exposures are the principal factors, with some cases of environmental exposure and exposure via food intake.

PCBS PCB data in the Region are very scarce and correspond to breast milk and adipose tissue levels reported by Chile. A total of 540 samples of human milk were analyzed: 18.3% was PCB positive with levels ranging from 0.09 to 84.93 mg/kg on a lipid basis (Tamayo et al. 1994). In contrast, a mean level of 54 ng/g fresh weight in samples of fatty tissue from persons living in an industrialized zone was also reported with the corresponding PCB TEQs values of 1.09 pg/g ww (Mariottini et al. 2000, 2002). These levels are almost an order of magnitude lower than values reported for relatively nonindustrialized areas in Italy (400 ng/g lipids).

IV. Discussion

The identification and quantification of POP sources and releases to the environment are poorly developed in the Region. Most of the results presented in this review refer to critical areas (hot spots), and to a lesser extent to stockpiles, which act as a major environmental source for the most common POPs. Industrial emissions and other specific sources, including the unintentional ones, are largely ignored, and no records are available to estimate the real threats that they may pose to humans and to the environment. Dioxin and furan emissions to the atmosphere were estimated using both the UNEP toolkit and CO_2 emissions. The value obtained for the whole region is ~700 g TEQ/year, and is in the same order of magnitude of values reported for European countries such as France, Belgium, and UK. Still, this estimation must be validated with experimental data.

The main PCB sources are improper use, disposal, and maintenance of the available stock. Considered in order of increasing priority, PCBs could be considered as the most important POP issue for the analyzed countries. The most relevant environmental problems with POPs are related to the improper use, disposal, and maintenance of the available stock. The discharge of untreated effluents throughout the whole region is recognized as a major input pathway of POPs into the environment.

The majority of the countries within the Region lack a routine monitoring program, and most of the available data were generated by induced monitoring rather than comprehensive programs in densely populated areas and along principal hydrographic basins. For this reason, the database concerning environmental levels strongly reflects bias toward freshwater ecosystems and does not address the current status and trends of vast areas of the region.

The environmental compartments most studied are aquatic animals, followed by sediments, water, and humans, with fewer data for air and soil. The majority of the published data corresponds to chlorinated pesticides and PAHs, with a few reports on PCBs and organic mercury. In general, in past decades, environmental pesticide levels of pesticides are decreasing, especially in foods, mainly because of legal restrictions or banning in the countries of the Region. However, the other POPs show clear contamination patterns in sediments, water, and biota in densely populated and industrialized areas.

An observed trend in the Region was a decline of pesticide levels during the last decade following banning of pesticides; however, it was recognized that a continuous monitoring effort is needed because there are large areas without information. Toxicological and ecotoxicological effects have been practically not assessed in the region, with only a few exceptions. Concern should be raised on the deferred effects of some POPs substances, and attention should be paid to chronic effects in humans in those situations where exposure is critical. With respect to environmental effects and responses, despite the attention to environmental levels in biota, few studies have addressed the effects of such exposure.

Studies on long range transport (LRT) at the regional level are practically nonexistent; only a few studies have been conducted in the Region, where several POPs such as PCBs, dioxins, furans, and other chlorinated compounds have been detected in remote areas (Andes Mountains and South Atlantic Ocean), showing the relevance of atmospheric and ocean transport. Taking into consideration the abundance and extensive area covered by the freshwater ecosystem, the LTR through this medium is very likely to occur at the highest altitude, although there are no experimental data to support this statement. Large rivers within the Region could be major transport pathways. In conclusion, the Region presents all the necessary conditions to allow the POP transport and spreading to remote areas, but there are few experimental data to evaluate the real contribution and importance of this mechanism to the present scenario of Regional POP assessment.

Summary

Persistent organic pollutants (POPs) are a global issue. The recently signed Stockholm POP convention requires information from signatory countries regarding sources and environmental levels. In eastern and western South American countries, this information is not always easily accessible, and therefore an effort toward collection of updated information is required. This review attempts to fulfill these requirements by analyzing the existing information regarding environmental levels of POPs in eight countries of South America.

A regional trend for environmental POP information is the uneven contribution among countries, which reflects the different patterns of economic, technical, and scientific development. In general terms, the available information is strongly biased toward those countries with scientists, and technical facilities for performing research in POP-related issues. Data related to environmental levels and spatial patterns principally come from the densely populated areas along major rivers such as the Amazon, Paraná, and Río de la Plata. The database is thus strongly biased toward freshwater environments to the detriment of coastal marine areas, which have received proportionally less attention.

POP monitoring in air is infrequent in the Region. Data represent a few geographical areas within the Region. PCB levels in air from some urban areas in Argentina, Brazil, and Chile are low to moderate (0.7–6.5 ng/m^3) but considerably higher than those reported for the remote Falkland Islands (Islas Malvinas) (5 pg/m^3). POP monitoring in soils is also limited within the Region. There are no regional monitoring programs, and most data refer to agricultural areas in Chile and urban hot spots in Brazil. Chilean soil data from agricultural areas indicate generally low levels of chlorinated pesticides in spite of a relatively high detection frequency.

Overall, several high pesticide levels for freshwater in the Region suggest a complex situation, but the narrow coverage of the data requires a cautious interpretation. Usually, only suspected contaminated ecosystems are monitored, and large-scale regional water monitoring programs have never been undertaken. The few PCB reports indicate generally low to moderate levels (7–22 ng/L), higher than recommended guidelines in urbanized estuaries and rivers such as the Río de la Plata (Argentina) and Biobio (Chile). Pesticide levels show high variability (0.6–14, 160 ng/L), reflecting distinct ecosystem conditions from less-impacted environments to severely polluted streams located in densely populated areas near Buenos Aires and Sao Paulo. Heptachlor, HCHs, aldrin, and DDTs are the most frequently reported pesticides in water, accounting for more than 60% of the total database.

Regional POP information for sediments is also dominated by chlorinated pesticides, but presents a more balanced contribution of PCBs with a few reports for PCDD/F. Overall, as observed for waters, sediment data indicate a complex situation in densely populated areas affected by urban-industrial inputs that have high POP levels. The most frequently reported POPs are DDTs, HCHs, PCBs, and heptachlors. The concentrations show a large variability, principally introduced by some highly contaminated sites in Argentina and Brazil with levels 4–5 orders of magnitude higher.

Aquatic organisms are by far the most studied organisms in the region, among them principally bivalves and fish. As observed for other environmental receptors, the regional distribution of data is uneven, heavily centered in coastal environments and in some countries (Argentina, Brazil, Chile, and Peru). The most comprehensive POP monitoring program in the South American coastal environment is the Mussel Watch. Among the POPs studied, PCBs predominate, followed by DDTs and chlordanes. Baseline PCB concentrations range from 200 to 700 μg/kg lipids in unpolluted sites, from 1,000 to 3,000 μg/kg in moderately contaminated sites, and from 4,000 to 13,000 μg/kg lipids in most affected bivalves from the Río de la Plata (Argentine side), Recife (Brazil), and Punta Arenas (Chile). DDT averages in bivalves are an order of magnitude lower than those of PCBs, below the 5 ppm guideline, and follow a similar spatial pattern.

Other organisms that have been analyzed for POPs in the Region are crabs (*Cyrtograpsus angulatus*), dolphins (*Pontoporia blainvillei*), and porpoises (*Phocoena spinipinnis*) from the Atlantic Argentine coast and continental shelf. PCBs show the highest levels (averages, 296, 1980 and 3300 μg/kg lipid in crabs, dolphins, and porpoises, respectively) closely followed by DDTs (170, 1670, and 4320 μg/kg lipid). PCDD/Fs analyzed in the omnivorous blue crab (*Callinectes sapidus*) collected in the Santos Bay area (Brazil) showed total TEQ values of 1.5 pg/g wet weight with the predominance of octachlor dioxins, followed by heptachlor congeners. POP data in crustaceans and gastropods from the Valdivia area (southern Chile) show generally low levels, basically of DDTs and DRINs (0.8–5 ng/g ww), whereas very few data from organisms collected in the Peruvian coast show low DDT (1–10 ng/g), and PCB levels (0.12–17.8 ng/g).

Acknowledgments

This review was partially supported by a GEF-UNEP Chemicals Project. The support of Chilean FONDECYT No 1010640 granted to R. Barra is greatly acknowledged.

References

Alvarez N (1998) Productos fitosanitarios: Su evolución en Argentina. Cámara Argentina de Sanidad Agropecuaria y Fertilizantes (CASAFE).

Astolfi E, Higa de Landoni J (1978) Residus des pesticides chlore dans le lait. Facultad de Medicina, Universidad del Salvador, Buenos Aires, Argentina.

Baker JI, Hites RA (2000) Is combustion the major source of polychlorinated dibenzo-*p*-dioxins and dibenzofurans to the environment? A mass balance investigation. Environ Sci Technol 34:2879–2886.

Ballschmitter K, Zell M (1980) Baseline studies of the global pollution. I. Occurrence of organohalogens in pristine European and Antarctic aquatic environments. Int J Environ Anal Chem 8:15–35.

Barra R, Cisternas M, Urrutia R, Pozo K, Pacheco P, Parra O, Focardi S (2001) First report on chlorinated pesticide deposition in a sediment core from a small lake in Central Chile. Chemosphere 45:749–757.

Barra R, Pozo K, Muñoz P, Salamanca M, Araneda A,Urrutia R, Focardi S (2004) PCBs in a dated sediment core of a high altitude lake in the Chilean altiplano: the Chungará lake. Fresenios Environ Bull 13:83–88.

Borrell A, Pastor T, Aguilar A, Corcuera J, Monzón F (1994) Contaminación por DDT y PCBs en Pontoporia Blainvillei de aguas Argentinas. Variación con la edad y el sexo. Anais do 2a Encontro sobre Coordenaçao de Manejo e Pesquisa da Franciscana Florianópolis, SC. EF/2/DT11.

Boroukhovitch A (1999) Current state of chlorine containing pesticides in Uruguay. Available at **http://irptc.unep.ch/pops/POPs_Inc/proceedings/Iguazu/URUGUAYE.html.**

Braga AM, Krauss T (2000) PCDD/F-concentrations in soil and cows' milk from a hexachlorocyclohexane contaminated area in Rio de Janeiro. Brazil. Organohalogen Compounds 46:354–357.

Cabrera E (1993) Determinación semicuantitativa de residuos de plaguicidas organoclorados (Aldrin) y carbamatos (Carbaryl) en Solanum Tuberosum: papa Tesis. Universidad Nacional Mayor de San Marcos, Lima.

Caviedes Vidal E (1998) Project Report Fisiología Ecológica de las Aves de San Luis. Línea Ecotoxicología: Búsqueda de bioindicadores de xenobióticos organoclorados en la Región centro-oeste.

Celeste MF, Cáceres O (1987) Resíduos de Pesticidas clorados em águas do reservatório Lobo (Broa) e seus tributários. Ciência Cultura 39:66–70.

CENMA (2001) Caracterización de los Bifenilos Policlorados (PCBs) en la atmósfera urbana de la Región Metropolitana de Chile.

CETESB (1997) Avaliação Preliminar De Área Contaminada Por Organoclorados (Depósitos De Agrotóxicos Do Município De Canoas (Rs): Fepam Cooperação Técnica Brasil-Alemanha. Outubro De 1997. Coordenação Inter-Projetos (Cip) Gtz/Fepam/ Feema/Cetesb.

CETESB (2000) Informação técnica no. 010/00/EEAS.

CETESB (2001) Companhia de tecnologia de saneamento ambiental, programa de controle de poluição. Sistema estuarino de santos e são vicente, agosto de 2001, disponível em cd e impresso.

CETESB (2002) Companhia de Tecnologia e Saneamento Basico de Sao Paulo. **www.cetesb.sp.gov.br** (accessed in June 2002).

Chagas CM, Lopes ME, Neves AA, De Queiroz JH, De Oliveira TT, Nagen TJ (1999) Determinação de compostos organoclorados presentes em rios da região de Viçosa, MG. Química Nova 22:506–508.

Colombo JC, Khalil MF, Arnac M, Horth A, Catoggio JA (1990) Distribution of chlorinated pesticides and individual polichlorinated biphenyls in biotic and abiotic compartiments of the Rio de La Plata. Environ Sci Technol 24(4):498–505.

CONAMA (Comisión Nacional de Medio Ambiente) (2001) PCBs en Chile. Diagnóstico Nacional de Contaminantes Orgánicos Persistentes. Documento de Trabajo No 2.

Corcuera J, Monzón F, Aguilar A, Borrell A, Raga AJ (2002) Life History Data, Organochlorine Pollutants and Parasites from Eight Burmeister's Popoises, *Phocoena spinipinnis*, Caught in Northern Argentine Waters. Special Issue on Phocoenids. Report of the International Whaling Commission, Cambridge.

Costa C (2000) Dias contados para o Ascarel Brasil. Energia 240:89–91.

Costa MA de C (1997) Estudo dos níveis ambientais de DDT em fauna e sedimentos na região da Baía de Todos os Santos. Dissertação de Mestrado em Química Analítica Instituto de Química, UFBA, Salvador, p. 65.

DINAMA-SOHMA-SHN (1998) Informe técnico. Impacto de Zonas Costeras. Módulo Salto-Concordia. Comisión Administradora del río Uruguay. Subcomisión de Contaminación.

DINAMA (Dirección Nacional de Medio Ambiente) (2000) Proceedings, UNEP Chemical Workshop on the Management of Polychlorinated Biphenyl, Dioxins and Furans. UNEP Chemicals, Montevideo, Uruguay, 19-22 Septiembre.

Farrington FW, Tripp BW (1995) NOAA Technical Memorandun NOS ORCA 95. International Mussel Watch Project. Initial Implementation Phase. Final Report. NOAA,

Focardi S, Fossi C, Leonzio C, Corsolini S, Parra O (1996) Persistent organochlorine residues in fish and water birds from the Biobio River, Chile. Environ Monit Assess 43(1):73–92.

García Fernandez JC, Marzi A, Casabella A, Roses O, Guatelli M, Villaamil E. (1979) Plaguicidas organoclorados en aguas de los ríos Paraná y Uruguay Ecotoxicología 1:51–78.

Gavilan JF, Barra R, Fossi MC, Casini S, Salinas G, Parra O, Focardi S (2001) Biochemical biomarkers in fish from different river systems reflect exposure to a variety of anthropogenic stressors. Bull Environ Contam Toxicol 66(4):476–483.

Gonzalez M, Miglioranza KSB, Gerpe MS, Menone ML, Lanfranchi AL, Aizpún de Moreno JE, Moreno VJ (2001) Acumulación de plaguicidas organoclorados (POC's) en vegetales comestibles cultivados en una huerta orgánica. IV Reunión Anual de SETAC Latinoamérica. Buenos Aires. Argentina. SQ6.

Grossi G, Lichtig J, Krauss P (1998) PCDD/F, PCB and PAH content of Brazilian compost. Chemosphere 37:2153–2160.

Higa de Landoni J (1978) Contaminación por Plaguicidas Clorados en la canasta familiar Argentina. Repercusión Biológica. Facultad de Medicina. Universidad del Salvador, Buenos Aires.

INDA (Instituto Nacional de Desarrollo Agroindustrial) (1987) Evaluación preliminar del grado de contaminación química en algunos alimentos. Lima, Peru.

INIA (1990) Fuentes de Contaminación con Residuos de Plaguicidas Organoclorados y Metales Pesados en Sectores Agrícolas, Regiones IV a XI. Ministerio de Agricultura, Chile.

Krauss P (2000) United Nations Environment Programme. Chemicals. Proceedings of UNEP Chemicals Workshop on the Management of Polychlorinated Biphenyls (PCBs) and Dioxins/Furans. 19–22 September, Montevideo, Uruguay.

Krauss P, Mahanke K, Freire L (1995) Determination of PCDD/F and PCB in forest soil from Brazil. Organohalogen Compounds 24:357–361.

Lamparelli MC, Kuhlmann ML, Carvalho MC, Salvador MEP, Souza RC, Botelho MJC, Costa MP, Martins MC, Carvalho PM, Araújo RPA, Buratini SV, Zanardi E, Sato MIZ, Roubicek DA, Valent GU, Rodrigues PF, Hachich EM, Bari MD, Curcio RLS Júnior APT, Lorenzetti ML, Truzzi AC, Pereira DN, Boldrini CV (1996) Avaliação do complexo Billings: Comunidades aquáticas, água e sedimento. Relatório técnico da Companhia de Saneamento Ambiental (CETESB) DAH.

Lara WH, Barreto HHC, Takahashi MY (1999) Resíduos de pesticidas clorados em frutas e verduras. Revista do Instituto Adolfo Lutz,

Lenardon A, Maitre de Hevia M, Enrique de Carbone S (1994) Organochlorine Pesticides in Argentinian butter. Sci Total Environ 144:273–277.

Lohmann R, Ockenden WA, Shears J, Jones KC (2001) Atmospheric distribution of polychlorinated dibenzo-*p*-dioxins, dibenzofurans (PCDD/Fs), and non-ortho biphenyls (PCBs) along a North–South Atlantic transect. Environ Sci Technol 35:4046–4053.

Lopes JLC, Casanoca IC, Figueiredo MCG, Nather FC Avelar WEP (1992) Anodontites trapesialis: a biological monitor of organochlorine pesticides. Arch Environ Contam Toxicol 23(3):351–354.

Mackay D (2001) Multimedia Environmental Models: The Fugacity Approach. Lewis, Boca Raton.

Maitre MI, de la Sierra P, Lenardon A, Enrique S, Marino F (1994) Pesticide residue levels in Argentinian pasteurized milk. Sci Total Environ 155:105–108.

Mandalakis M, Stephanou EG (2002) Polychlorinated biphenyl associated with fine particles (PM 2.5) in the urban environment of Chile: concentration levels and sampling volatilization losses. Environ Toxicol Chem 21(11):2270–2275.

Mariottini M, Aurigi S, Focardi S (2000) Congener profile and toxicity assessment of polychlorinated biphenyls in human adipose tissue of Italians and Chileans. Microchem J 67(1–3):63–71.

Mariottini M, Guerranti C, Aurigi S, Corsi I, Focardi S (2002) Pesticides and polychlorinated biphenyl residues in human adipose tissue. Bull Environ Contam Toxicol 68:72–78.

MDIC (Ministério do Desenvolvimento da Indústria e Comércio, Brazil). **http://aliceweb.mdic.gov.br** (accessed in August 2002).

Menone ML, Aizpún de Moreno JE, Moreno AL, Lanfranchi TL, Metcalfe TL, Metcalfe CD (2001) Organochlorine pesticides and PCBs in a Southern Atlantic Coastal Lagoon Watershed, Argentina. Arch Environ Contam Toxicol 40:355–362.

Ministry of Health Chile (1994) Analysis of Organochlorine Pesticides in Foods, Years 1983–1993. Public Health Institute of Chile, Subdepartamento Bromatología.

Muñoz J, Becker PH (1999) The kelp gull as bioindicator of environmental chemicals in the Magellan region. A comparison with other coastal sites in Chile. Sci Mar 63(suppl 1):495–502.

MVOTMA-DINAMA-UNEP (2002) Inventario Nacional de liberaciones de dioxinas y furanos, Uruguay—2000. Ministerio de Vivienda, Ordenamiento Territorial y Medio Ambiente-Dirección Nacional de Medio Ambiente-UNEP, 54 pp.

Natale OE, Gomez CE, Pechen de D'Angelo AM, Soria CA (1988) In: Abbou R (ed) Waterborne Pesticides in the Negro River Basin (Argentina). Hazardous Waste: Detection, Control, Treatment: Proceedings of the World Conference on Hazardous Waste, Budapest, Hungary, Oct. 25–31, 1987, Part A. Elsevier, Amsterdam, pp 879–907.

Ockenden W, Lohmann R, Shears J, Jones KC (2001) The significance of PCBs in the atmosphere of the southern hemisphere. Environ Sci Pollut Res 8:189–194.

Oliveira RM, Brilhante OM (1996) Hexachlorocyclohexane contamination in an urban area of Rio de Janeiro, Brazil. Environ Int 22:289–294.

Palma-Fleming H, Espinoza O, Gutierrez E, Pino M (1998) Organochlorine pesticides in sediment of the Valdivia river estuary in Chile. Bol Soc Ch Quim 43(4):435–445.

Rodriguez Girault ME, Alvarez G, Ridolfi A, Rovenna A, Mirson D, Villaamil E, Lopez CM, Roses O (2001) Plaguicidas Organoclorados en Leches Infantiles. XII Congreso Argentino de Toxicología, VA-8, Rosario, Argentina, p. 30.

Rovedatti MG, Castañe PM, Topalian ML, Salibian A (2001) Monitoring of organochlorine and organophosphorus pesticides in the water of the Reconquista River (Buenos Aires, Argentina). Water Res 35:3457–3461.

Sericano J, Wade T, Jackson T, Brooks J, Tripp B, Farrington J, Mee L, Readman J, Villeneuve JP, Goldberg E (1995) Trace organic contamination in the Americas: an overview of the US National Status and Trends and the International Mussel Watch Programmes. Mar Pollut Bull 31:214–225.

Tamayo C, Matus MN, Montes SC, Cristi VR (1994) Polychlorinated biphenyls (PCBs) determination in human milk samples collected in three provinces of the tenth region of Chile 1990. Rev Med Chile 122(7):746–753.

Tavares TM, Bereta M, Costa MC (1999) Ratio of DDT/DDE in the All Saints Bay, Brazil and its use in environmental management Chemosphere 38(6):1446–1553.

Telles DL (2001) Inseticidas organoclorados e bifenilos policlorados (PCBs) na região estuarina de Itamartacá/PE: aspectos analíticos e ambientais. Tese de Doutorado Recife-PE,

Topalián ML, Castañé PM, Rovedatti MG (1996) Dos años de monitoreo regular de plaguicidas en el agua del Río Reconquista. X Congreso Argentino de Toxicología, vol 34. Buenos Aires, Argentina, p. 35.

Torres JPM, Malm O, Vieira EDR, Japenga J, Koopmans GF (2002a) Organic micropollutants on river sediments from Rio de Janeiro state, Southeast Brazil. Cadernos de Saúde Pública Rio de Janeiro 18(2):277–488.

Torres JPM, Pfeiffer WC, Markowitz S, Pause R, Malm O, Japenga J (2002b) Dichlorodiphenyltrichloroethane in soil, river, sediment and fish in the Amazon in Brazil. Environ Res 88(2):134–139.

UNEP (1999) Dioxin and Furan Inventories, National and Regional Emissions of PCDD/PCDF. United Nations Environment Programme, UNEP Chemicals, Geneva, Switzerland.

UNEP (2003a) Regionally Based Assessment of Persistent Toxic Substances. Global Report, UNEP Chemicals. Geneva, Switzerland.

UNEP (2003b) Standardized Toolkit for Identification and Quantification of Dioxin and Furan Releases. UNEP Chemicals, Geneva, Switzerland.

Vilar de Saráchaga D (1997) Perfil nacional para la gestión de sustancias químicas. Foro Intergubernamental de Seguridad Química. United Nations Institute for Training and Research (UNITAR).

Villaamil EC (2000) Residuos de plaguicidas organoclorados en alimentos de consumo habitual en la ciudad de Buenos Aires. Tesis doctoral, Toxicología y Química Legal, FFyB, UBA.

Villaamil EC, Ridolfi A, Ravenna A, Pongelli V, Roses O (1999a) Investigación de residuos de plaguicidas organoclorados en alimentos no grasos de consumo habitual en Buenos Aires. XI Congreso Argentino de Toxicología, AN16. Buenos Aires, Argentina, p. 83.

Villaamil EC, Ridolfi A, Ravenna A, Paonessa A, Roses O (1999b) Investigación de residuos de plaguicidas organoclorados en alimentos grasos de consumo habitual en Buenos Aires. XI Congreso Argentino de Toxicología, Buenos Aires, Argentina.

Wilcke W, Lilienfein J, Do Carmo L, Bayreuth B (1999) Contamination of highly weathered urban soils in Uberlandia. J Plant Nutr Soil Sci 162:539–548.

Whelan J, Hunt J (1983) Volatile C1-C8 organic compounds in sediments from the Peru upwelling region. Org Geochem 5:13–28.
World Resources (2000) 2000–2001: People and Ecosystems: The Fraying Web of Life. United Nations Development Program, United Nations Environment Program, World Bank, World Resources Institute. Elsevier, Amsterdam.

Manuscript received May 11, 2004; accepted May 26, 2004.

Rev Environ Contam Toxicol 185:35–65

Ecotoxicological Assessment of the Highly Polluted Reconquista River of Argentina

Alfredo Salibián

Contents

I. Introduction

The aquatic environment is the final receptor of all human-made or natural contaminants. The situation is particularly critical near large human conglomerations of Latin America that have grown disproportionately and without any planning. Extremely high urban and industrial concentrations are associated with environmental pollution processes that, in turn, cause degradation of air, water, and soil, as well as of other natural resources and of the quality of life of people living in those places (Borthagaray et al. 2001; Finkelman 1996; Hajek 1995; Joyce 1997).

The expected trend of urban population growth in Argentina shows that the percentage of urban population will increase up to 96% by the year 2025; similar tendencies are predicted for all countries of the region (Escalona and Winchester 1994). In fact, the population census for Argentina indicated that urban population grew from 87.2% in 1991 to 89.3% in 2001 (INDEC 2002).

Those degrees of urbanization in Latin America are close to or even higher than those of many industrialized countries; however, although in those countries almost 100% of the urban population has access to safe drinking water and sanitation services, this percentage is considerably lower for the South American

---

Communicated by Lilia Albert.

Alfredo Salibián (✉)
Comisión de Investigaciones Científicas de la Provincia de Buenos Aires, and Programa de Ecofisiología Aplicada (PRODEA), Departamento de Ciencias Básicas, Universidad Nacional de Luján, B6700ZBA-Luján, Argentina.

Dedicated to the memory of Professor Mirta L. Topalián.

urban communities; the particular situation for the population settled in the Reconquista River's basin appears to be critical (Saltiel 1997a).

In addition, most of the large South American cities are surrounded by shantytowns of poor and jobless people who migrated from the countryside to seek a better standard of life. The speed of this demographic process has increased in particular in the past two decades after the economic models that have been imposed in the region. It is worth mentioning that the aforementioned economic policies also gave rise to the reduction or weakening of the controlling and regulating capacity of the government regarding the environmental problems; therefore, the situation has worsened over time.

More recently, the picture is growing even worse because of an urbanization process taking place in reverse, revealing a centrifugal movement of people who migrate from large urban centers to gated communities located in the surroundings of the big cities, thus occupying lands of high productive capacity and low levels of pollution (Mateucci et al. 1999). This process may explain the fact that the population of Buenos Aires city, the largest of the country, declined between 1991 and 2001 by 6.4%.

In Argentina, freshwater pollution is critical in urban and near-urban water bodies close to the large urban centers, particularly in the estuary of Río de la Plata (in Spanish, river = *río*). It is a kind of a collecting "funnel" of all pollutants brought by the Paraná and Uruguay Rivers that carry away the agricultural, industrial and domestic wastes from several neighboring countries of the region; it also receives wastes from the large San Nicolás–Río Santiago industrial axis, which includes the city of Buenos Aires and the Great Buenos Aires. It is approximately 500 km long and receives contaminant inputs of high complexity and chemical variability, coming from the atmosphere and by direct point discharges (Brailovsky and Foguelman 1992; Tudino 2001).

Several authors have pointed out that providing freshwater in the region will become a critical issue in the near future (Fernández Cirelli 1998). In 2000, the average renewable water resources in Argentina were 9721 m^3/capita/yr, the lowest in the region, with 75% of the surface freshwater withdrawals for agricultural use, 9% for industry, and 16% for domestic use. In addition, freshwater withdrawals are approximately 950 m^3/capita/yr, 127 of which are groundwater (UNDP et al. 2000).

At present, most of the water bodies of Argentina are affected in some degree by pollution processes (Loez and Topalián 1999). Foguelman and González Urda (1994) estimated that in Argentina only 10% of the industrial wastes are retained. In other words, 90% of the wastes produced in industrial activities go, without treatment, directly to water bodies.

The rivers and streams near the city of Buenos Aires and its outskirts receive some 300,000 t yr^{-1} of toxic mud, 500,000 t yr^{-1} of diluted solvents, and 500,000 t yr^{-1} of effluents with elevated quantities of heavy metals. In addition, there is unequivocal evidence showing the existence of groundwater pollution.

The situation in the majority of the large rivers in Argentina is very serious due to the amount and range of chemical pollutants, principally as a consequence of industrial development with an inadequate regulatory framework and a deficit of decades in matters of sanitary substructure and waste treatment. In recent years, the intensity of the environmental stress conditions have shown, in some cases, incipient signs of recovery, which was interpreted as due to acute economic recession in the country, reflected by an important reduction in industrial activity.

Pollution levels increase in these large rivers in a sustained way in time, particularly where pesticides, heavy metals, and hydrocarbons and other products derived from oil, present either in the biota, in solution, or in sediments, reach values sometimes much higher than those established by the Argentinean Law of Hazardous Wastes (No. 24051) for the different uses of water (Bilos et al. 1998; Catoggio 1990; Colombo et al. 1989, 1990, 1997; Comisión Administradora del Río de la Plata 1989; Ronco et al. 1996; Verrengia-Guerrero and Kersten 1998; Villar et al. 1998).

The history of regular and integral freshwater quality monitoring in Argentina is limited. Where measurements were made, these were monitored at a limited number of sites restricted to pollution sources, without a prolonged temporal sequence of samplings. Monitoring programs relied mainly on water chemistry and bacteriology, with measurements carried out for only the main variables, those required for the determination of particular water quality indexes. The multidisciplinary approach considering simultaneous evaluation of a number of factors and processes that, in an integrative picture, may determine its characteristics, was poorly developed or lacking. The use of biota for monitoring the quality of aquatic environments has been relatively uncommon compared to abiotic variables (Topalián et al. 2001).

II. Reconquista River Description

The Reconquista River is a typical lowland watercourse situated in the northwest of Buenos Aires Province. Located in a temperate subtropical region, it originates in the confluence of the Durazno, La Choza, and La Horqueta streams and finally joins the Luján River that flows into an international river (Río de la Plata) (Fig. 1), which is part of the second largest hydrogeographic system of South America, after the Amazon, and the fifth largest in the world.

The length of the river is approximately 55 km with a low flow (70,000–1,700,000 $m^3 d^{-1}$). Its width varies from 4 to 25 m and its depth fluctuates between 0.5 and 2.5 m, with an approximate downgrade slope of 0.5 $m\,km^{-1}$.

The river receives the output of 80 small tributaries; one of them, the Morón creek, should be highlighted as it marks the limit between the medium and the lower sections of the river. It is roughly 16 km long, partially tubed, with a mean flow of 0.9 $m^3 sec^{-1}$. During the drought season the entire stream is composed of sewage and industrial wastes and is known as an "open sewer" (Kuczynski 1991a, 1994).

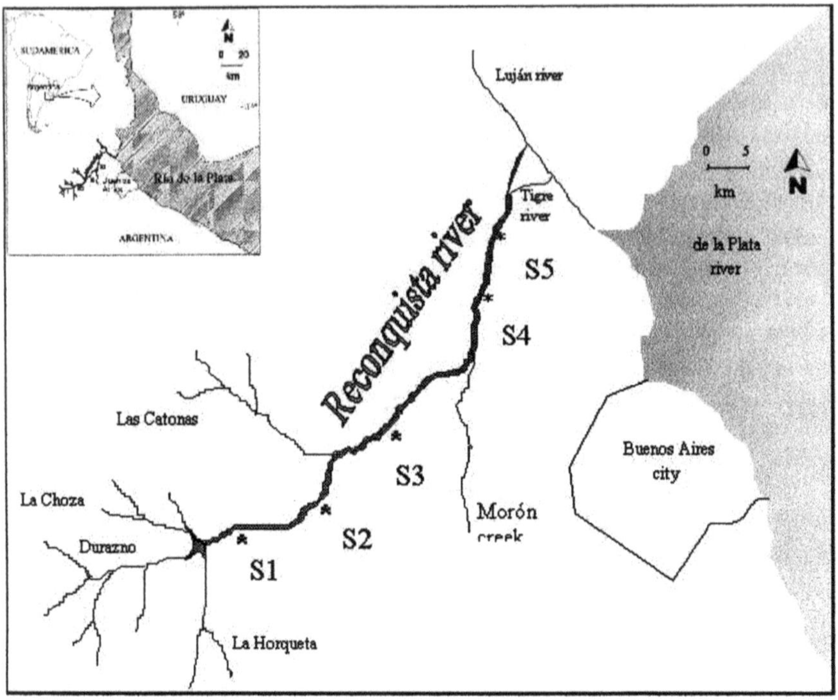

Fig. 1. Reconquista River, showing positions of sampling sites: S1, Cascallares; S2, Paso del Rey; S3, Gorriti; S4, San Martín; S5, Bancalari

The watershed of the river covers about 1700 km^2. It includes three areas: the upper plain, 720 km^2, is mainly a farming and animal husbandry area while the other two plains are either urbanized or soon will be. The latter includes green and public areas such as the one used by CEAMSE (State Society of Ecological Coordination of the Metropolitan Area) as an open-air landfill zone to hold unclassified solid urban wastes from the city of Buenos Aires and its surroundings (Zalazar 1996). These processes of land occupation, which take place simultaneously with others, modify in a permanent way the original characteristics of the watershed.

Watershed rainfall is quite irregular in frequency and magnitude, without a defined dry season throughout the year. Mean annual rainfall was 1034 mm during the studied period.

At present (2001 Population Census), approximately 3.5 million people are settled in the highly stressed basin of the Reconquista River. Between 1991 and 2001, that population increased by 8.9%. These figures correspond to approximately 10% of the total population of the country and to 32% of the Buenos Aires city plus its metropolitan area. It is interesting that the increase of settling rate on the river's basin was slightly lower than the growth rate of the total population of the country, which was, for the same period, 11.1% (INDEC 2002).

III. Pollution of the Reconquista River

It is accepted that the human colonization process of the river's basin began in the middle of the 16th century. A sustained industrial activity was recorded by the end of the 18th century. At that time, a number of sawmills were established near the mouth of the river (Kuczynski 1993).

Other factors that have had to do with the human settlement, different from those related to industrial activities, are worth mentioning briefly. In 1871 an epidemic of yellow fever affected Buenos Aires, causing a massive migration of people that moved toward the northeast close to the mouth of the Reconquista River area. Earlier, in 1567 cholera epidemics moved in from Paraguay.

The strongest industrial development of the country began at the first third of the 1900s. As early as the 1950s, people complained about the river's pollution.

The Reconquista River is an example of the adverse impact of human activity on an urban related waterway. It is the second most polluted river of Argentina, with a great variety of industrial activities settled in its basin (textiles, tanneries, dairy and meat processing, chemical, metallurgic, electroplating, *etc.*); 85% of these industries are concentrated in five municipalities. In all, 10,000 plants, most of them located on the margins of the river (Alsina and Herrero 2000), discharge their untreated effluents into the river and use large quantities of water in processing, cooling, and cleaning. Approximately 20% of these industries discharge a total biological oxygen demand (BOD) load of approximately 150,000 kg d^{-1}, equivalent to an organic loading capacity of 2.5 million population.

The main course of the river and its tributaries receive running water from different origins, (landfills, agricultural, domestic, industrial, *etc.*) (Saltiel 1997b). Furthermore, the water that flows from the urban margins, and from industrial and agriculture and cattle farming activities, must be added. This water carries away a huge and diverse mass of wastes and chemical substances, exceeding in almost all cases the self-depuration capacity of the river. This situation is serious, not only because of the increasing amount of anthropogenic wastes but also because of their alarming variety. Thus, the water of the river becomes an extremely complex mixture of chemicals that, after physical, chemical, and biological alterations and interactions, both synergistic and antagonistic, may become a factor of high toxicological and ecotoxicological risks.

It must be kept in mind that the drainage system is not large enough; therefore, deposition of domestic effluents is inappropriate and further degrades the quality of water. Industrial activity, together with demographic growth, has contributed to the deterioration not only of the surface water but also the groundwater of the basin, as well as pollution of the adjacent soil (Pereyra and Tchilinguirian 2003). The groundwater of the area is a main source of running water from which inhabitants with no service of piped potable water draw their supply.

In past years, several floods were reported in the watershed area of the river; the most recent ones determined the scattering of different contaminants in the soil and in the water. Marbán et al. (1999) have shown that soils near the river affected by those floodings were heavily contaminated with Pb, Hg, and Cd. As a consequence, the Reconquista River enhances the serious contamination problem of the major basin of Río de la Plata, whose southern riverside is the most important source of freshwater for human use in the cities of Buenos Aires and La Plata and their outskirts, with a population of about 12 million inhabitants.

In spite of the fact that the Río de la Plata has a considerably higher capacity of self-depuration by dispersion and dilution because of its large dimensions, Villar et al. (1998) determined that Cd, Zn, Pb, and Cr concentrations in the water were higher than the maximum permitted quantities in sites located a short distance downstream from the outlet of the Reconquista River. In addition, there is evidence that concentrations of heavy metals in the sediments of coastal areas of the Río de la Plata are very high (Kreimer et al. 1996).

The aforementioned description, typical of a difficult and serious situation, confirmed that every particular environmental process is in fact a part of a larger system and that what happens in its diverse subsystems strongly affects the other subsystems.

IV. National University of Luján Studies

Studies on the quality of surface waters of the Reconquista River, during the 15 years between 1985 and 2001 is the longest and most complete regular monitoring of continental waters in Argentina in the past quarter-century (Salibián 1999). Following our studies, other authors have contributed additional information on the water quality by determining some physicochemical parameters and the heavy metal content of the sediments of Durazno and La Choza, two of the creeks that give rise to the Reconquista River (Tudino 2001).

To monitor water quality in the main course of the Reconquista River, research activities have concentrated on the following:

- Measurement of physical and chemical variables, about 30 parameters for each water sample
- Determination of water quality indexes based on the chemical profiles
- Determination of biological parameters (phytoplankton and zooplankton abundance, diversity and community structure, microbiology)
- Acute and prolonged *in situ* and laboratory toxicity tests with native and standardized producer and consumer species as sentinel organisms

When they agreed, various numerical and statistical procedures were applied to the different sets of data.

Studies were carried out on samples taken monthly in five sites, covering the entire length of the main course of the river: Cascallares (S1), Paso del Rey (S2), Gorriti (S3), San Martín (S4), and Bancalari (S5) (see Fig. 1).

Cascallares is located close to the mouth of the river (5 km) and was identified as the "control" site because of its low pollution rate (Loez and Salibián 1990), affected mainly by agricultural runoff and livestock watering. Paso del Rey (10 km) and Gorriti (20 km) are located in the rural–urban transition, with increasing loads from industrial and household waste discharges. San Martín (S4) (38 km downstream) was considered to be a key site to evaluate the impact of toxic wastes of the Morón creek, which flows into the Reconquista just above this point. Bancalari (S5) (45 km) was the site that showed the heaviest deterioration; it lies close to the outlet in the Luján River, but is far from the influence of the Rio de la Plata tides which might alter the composition of water samples. At this point, the water is always blackish-brown, with heavy foam on the surface and evidence of the presence of petroleum and its waste products.

The area of the river whose margins are used by the CEAMSE is located between San Martín and Bancalari. In this respect, it is interesting to mention the results reported by Faggi et al., (1999) after detailed studies on the aquatic and riparian vegetation of the river. As they moved from Cascallares (S1) through Bancalari (S5) during sampling, they reported a significant decrease in the number of native species and a parallel increase in alien species. They concluded that their findings cannot be explained only as a secondary consequence of the pollution and, alternatively, postulated that the interpretation of those changes in the riparian flora profile might be related to anthropogenic factors such as the impact of urbanization processes and mechanical modifications occurring in the margins of the river as a result of changes in land use.

It is necessary to highlight that the water quality measurements that focus on levels of contaminants are useful but do not directly tell us how complex and poorly characterized mixtures of a large number of chemicals pollute water and affect adversely freshwater ecosystems. The evaluation of river water quality by physicochemical parameters provides a temporary approach and reveals only the presence of the substances that are being determined. In systems of mixed and complex pollution, such parameters are not sufficient to infer joint effects on the aquatic biota.

Toxic substances may interact among themselves, and with natural substances, particles, and organisms, thus varying their toxicity and their bioavailability. Therefore, data related to the structure and dynamics of the communities such as phytoplankton and zooplankton are particularly important because they can indicate environmental quality alterations by their particular sensitivity to the effects of changes in nutrient concentrations and to the presence of toxics and xenobiotics.

Therefore, an integrated protocol of assessment of water pollution is a basic requirement for the design of its control strategies. In our case, the information acquired and the integration of both abiotic and biotic features of the water resulted in a more comprehensive understanding of the dynamics of the pollution status of the river, characterizing it both spatially and temporarily. In addition, we determined different chemical pollution indexes as complementary diagnostic information referring to the quality of the river water.

More recently, a number of statistical procedures (cluster analysis and ordination by PCA (Principal Components Analysis) stepwise discriminant analysis, and multiple discriminant analysis of variance) (Hair et al. 1995; Legèndre and Legèndre 1979; Plá 1986) have been recommended as suitable tools for the study of aquatic communities and to correlate them with the profile of the physicochemical parameters in their particular environment. By means of this methodology, we were able to detect phytoplankton and zooplankton associations that may be representative of particular environmental conditions (Olguín Salinas et al. 1999; Rovedatti et al. 2000; Topalián et al. 1999a).

The toxicity bioassays also contribute to this integrative approach concerning their acute response or through the changes that can be seen in particular bioindicators of different freshwater species used as test organisms under chronic or sublethal conditions. This type of approach was recently applied to ecotoxicological assessment of the quality of the river water, showing that the integrative complementation of different chemical and biological analytical information with appropriate mathematical methods of analysis may be more precise tools (Olguín et al. 2004).

Our team conducted this study that supplied a reliable basis, particularly useful to (a) determine the spatiotemporal quality or deterioration of the waters, (b) monitor the impact of remediation programs, (c) establish regulatory needs, and (d) design predictive models of what the future would be like relative to the quality of the river water under different environmental conditions.

V. Analytical Portrait of the Reconquista River

To shorten the text, the results are shown in a general way in selected figures and tables that partially present the data obtained in our studies. In each case, the specific sources of information were indicated so that the reader can obtain a detailed description of the applied analytical methods as well as of the results reached.

A. Abiotic Parameters

Basic Physiocochemical Parameters Most of the physical and chemical parameters (Table 1) showed changes downstream, particularly acutely, and in some cases abruptly, after the discharge of the highly polluted Morón stream, indicating significant deterioration of the water quality in the zone between S4 and S5 (Castañé et al., in preparation).

Water temperature did not differ among sites, suggesting that thermal pollution was not a common event in the river; it changed in concordance with seasonal variations, oscillating between 6° and 10 °C in winter and between 30° and 31 °C in summer. The pH along the river was mostly in the alkaline zone (7.4–10.0), characteristicly a steady high-to-low gradient from the source to the mouth, which may result from to the calcareous substratum of the river.

The majority of the remaining chemical parameters showed a clear tendency to increase toward the outlet of the river, with important jumps in the San Martín–Bancalari area.

The spatial variation of the dissolved oxygen (DO) was very important, from 7–8 $mg\,L^{-1}$ in S1 down to 0–0.3 $mg\,L^{-1}$ in S5; thus, the water condition in Bancalari was permanent anoxia. The DO in less-polluted points (e.g., S1) increased when the algal density was augmented by the blooms.

Table 1. Physicochemical parameters of water samples of the Reconquista River water taken fron Cascallares (S1), Paso del Rey (S2), Gorriti (S3), San Martín (S4), and Bancalari (S5), 1994–1995 monitoring.

Parameter	Year	S1	S2	S3	S4	S5
Temperature (°C)	1994	19.1 (1.6)	19.6 (1.6)	20.3 (1.5)	21.6 (1.7)	21.3 (1.7)
	1995	19.6 (1.6)	19.9 (1.7)	20.5 (1.5)	21.0 (1.6)	20.7 (1.5)
Conductivity ($\mu S\ cm^{-1}$)	1994	992 (96)	1007 (92)	933 (83)	1268 (98)	1279 (104)
	1995	842 (95)	891 (92)	883 (74)	1199 (108)	1217 (120)
Alkalinity (mg $CaCO_3$ L^{-1})	1994	947 (69)	968 (77)	971 (69)	1026 (58)	997 (55)
	1995	767 (90)	828 (88)	870 (65)	900 (79)	890 (84)
Turbidity (NTU)	1994	120 (18)	106 (19)	116 (49)	100 (14)	89 (11)
	1995	87 (20)	79 (16)	63 (12)	75 (4)	88 (7)
Nitrites (mg NO_2^--N L^{-1})	1994	0.12 (0.02)	0.24 (0.03)	0.25 (0.06)	0.14 (0.06)	0.17 (0.07)
	1995	0.15 (0.07)	0.35 (0.08)	0.32 (0.05)	0.08 (0.02)	0.06 (0.01)
Nitrates (mg NO_3^--N L^{-1})	1994	1.65 (0.38)	2.43 (0.41)	1.85 (0.32)	1.34 (0.34)	1.43 (0.41)
	1995	0.66 (0.11)	1.64 (0.22)	1.06 (0.19)	0.79 (0.19)	0.84 (0.29)
Ammonium* (mg NH_4^+-N L^{-1})	1994	0.7 (0.1)	3.8 (0.9)	7.9 (1.2)	11.2 (1.2)	11.1 (1.5)
	1995	0.7 (0.05)	4.3 (1.0)	7.7 (1.0)	11.3 (1.6)	11.1 (1.6)
Phosphates* (mg PO_4^{3-} L^{-1})	1994	1.65 (0.28)	2.74 (0.43)	3.85 (0.39)	4.78 (0.47)	5.09 (0.58)
	1995	1.66 (0.16)	2.98 (0.42)	4.00 (0.35)	4.83 (0.47)	5.19 (0.63)
Hardness* (mmol $CaCO_3$ L^{-1})	1994	1.3 (0.1)	1.3 (0.0)	1.4 (0.1)	1.8 (0.1)	1.9 (0.1)
	1995	1.2 (0.08)	1.4 (0.09)	1.4 (0.06)	1.9 (0.1)	1.9 (0.1)
Chlorides* (mg L^{-1})	1994	63.9 (6.3)	61.9 (5.2)	52.3 (5.0)	111.2 (6.7)	118.3 (11.2)
	1995	60.0 (9.6)	58.7 (6.8)	53.7 (5.1)	121.7 (14.8)	127.7 (18.8)
Phenols* (mg L^{-1})	1994	0.54 (0.05)	0.48 (0.05)	0.47 (0.05)	0.93 (0.09)	0.99 (0.14)
	1995	0.61 (0.05)	0.57 (0.04)	0.52 (0.06)	1.49 (0.32)	1.21 (0.30)

(continued)

Table 1. (Continued)

Parameter	Year	S1	S2	S3	S4	S5
BOD* (mg O_2 L^{-1})	1994	6.4 (1.1)	11.7 (1.6)	12.7 (1.3)	43.5 (7.4)	36.3 (6.1)
	1995	5.3 (1.0)	11.2 (2.4)	12.9 (2.1)	48.6 (9.0)	49.3 (7.5)
COD* (mg O_2 L^{-1})	1994	72.2 (5.7)	75.2 (5.7)	76.2 (5.1)	194.6 (25.7)	164.7 (26.0)
	1995	59.1 (3.2)	63.4 (5.0)	69.7 (5.4)	171.2 (27.1)	192.9 (27.2)
DO* (mg O_2 L^{-1})	1994	7.3 (0.7)	6.5 (0.9)	2.7 (0.6)	0.5 (0.2)	0.4 (0.2)
	1995	7.3 (0.5)	6.5 (0.5)	3.9 (0.4)	0.7 (0.3)	0.3 (0.2)
Chlorophyll *a* (μg L^{-1})	1994	34.1 (8.7)	30.4 (7.9)	30.0 (7.9)	31.8 (11.1)	29.2 (8.1)
	1995	19.6 (8.1)	24.7 (7.9)	29.0 (14.1)	35.3 (13.8)	21.9 (8.1)

BOD: biological oxygen demand; COD: chemical oxygen demand; DO: dissolved oxygen.
*Significant differences ($P < 0.05$) between S1, and S2 with respect to S4 and S5.

As expected, values of conductivity, BOD, and chemical oxygen demand (COD) also increased in the direction S1 to S5; at S4–S5 there were always registered values 2 to 20 times higher than those of S1–S3. The hardness was considerably uniform except in S4–S5 samples, where it reached values approximately 40% higher than in S1–S3. The COD/BOD ratios oscillated between 11 (in S1) and 4 (in S4–S5), suggesting the presence of important amounts of nonbiodegradable organic matter along the river.

Physicochemical parameters particularly important from the toxicological and ecotoxicological aspects, indicators of domestic sewage and municipal wastes, such as chloride, phosphates, inorganic N compounds (NH_4^+, NO_2^-, NO_3^-), and phenols, were found in all samples. The data showed a clear-cut trend of increasing concentration toward the mouth of the river, reaching values always well over the maximum permitted quantities (MPQs), the increases being particularly accentuated between S4 and S5.

For instance, N-NH_4^+ was 15 times higher in S5 compared with S1. When the concentration of the same cation is expressed as a percentage of total dissolved inorganic nitrogen ($NH_4^+ + NO_3^- + NO_2^-$), in S4–S5 it represented 80%–93%. Similar tendency to increase suddenly after S4 was recorded with phosphates, chlorides, and phenols (Topalián et al. 1999a,b). These high nutrient inputs are mostly related to sewage effluents and secondarily to industrial discharges. Consistently, BOD values also increased downriver. Within this framework is underlined the presence of phenols, always in concentrations 60 to 150 fold higher than the maximum permitted quantities.

Abundant foam, attributed to domestic or industrial drainage of surfactants, was frequently observed, which in aquatic media reduces transparency and gas exchange at the air–water interface.

Heavy metals in water and in sediments It can be said that heavy metals are permanently present in the river. At the same time, it was noted that the values suffer randomly sudden important increases. These findings were not restricted to any particular area of the river, which was interpreted as evidence that it is used as an illegal repository site for industrial and domestic wastes. In agreement with this observation is the fact that rainfall did not significantly change the profile just described.

Because of the proximity of Buenos Aires and its outskirts, it is plausible to postulate that a fraction of the detected metals in the water of the river may also have atmospheric origin by deposition from different points of the industrial belt of the city.

The total heavy metal concentrations found throughout the sampling period exceeded widely the MPQs established by Argentine law for protection of freshwater life, being, for example, 4 (As), 40,000 (Cd), 150 (Cr), 65 (Cu), and 23 (Zn) times higher, respectively, than those limits. In Fig. 2 are shown data corresponding to three heavy metals found in surface water from Cascallares, San Martín and Bancalari, sampled between 1994 and 1996 (Topalián et al. 1990, 1999b).

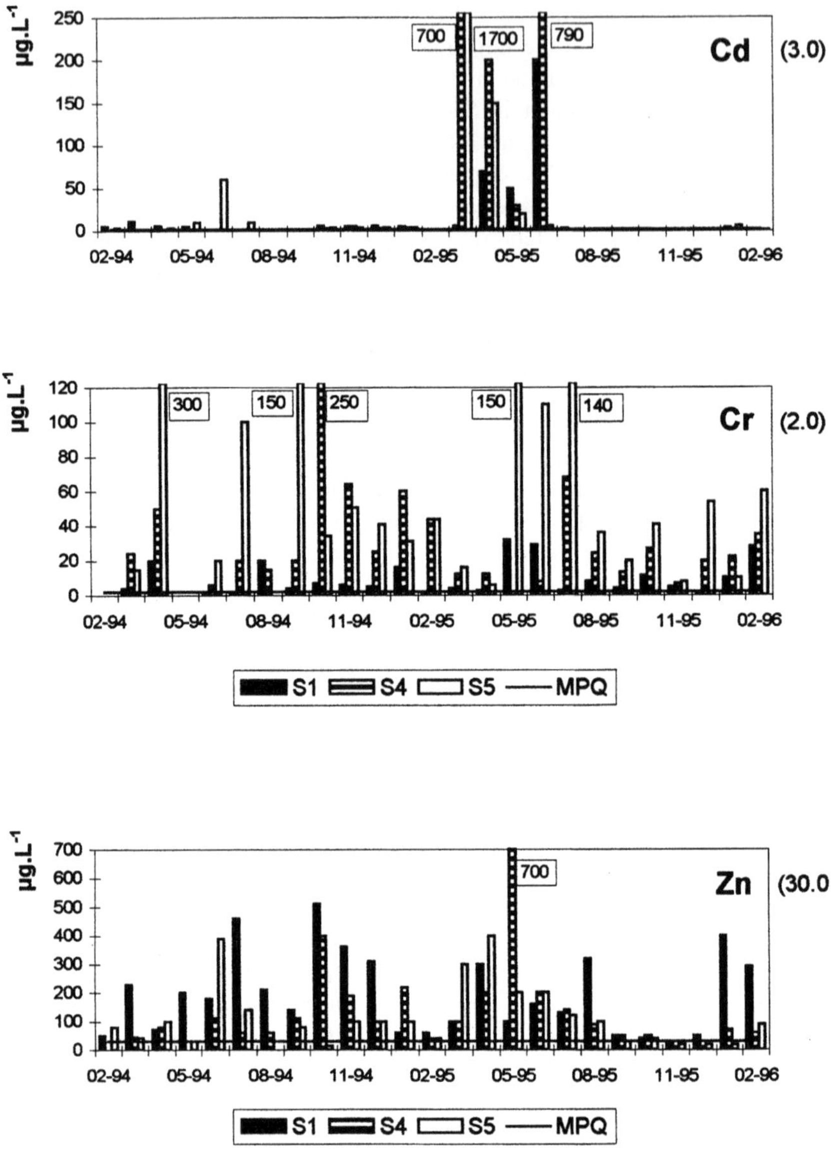

Fig. 2. Concentration of Cd, Cr, and Zn in the Reconquista River. In parenthesis are MPQ values. *Source*: Topalián et al. (1996)

It is known that particles of reactive metals released to bodies of water are likely to accumulate in sediments. Seasonal variability in the temperature of the river water may have led to temporal variability in the flux of deleterious chemicals from the sediments, acting as a removal factor from the overlying water column (Castañé et al. 1998a). In that respect, it was reported that extremely large amounts of heavy metals were found in sediment samples taken in San Martín (S4), in particular, Cr (1923 mg L^{-1}) and Zn (967 mg L^{-1}); other metals (Cu, Cd, Pb, Mn) were also found at high concentrations associated with considerable quantities of humic and fulvic acids (Iorio et al. 1997; Saltiel and Romano 1997; Tudino 2001).

Insecticides in Water Large numbers and quantities of pesticides are used in Argentina for agricultural and public health purposes. It is well known that their application may be a contaminant source for the aquatic environment, in particular in developing countries where their use is estimated at 20% of the total amount of pesticides used in the rest of the world.

Most of the applied pesticides are subject to many transport and conversion processes. Thus, they do not remain at their target site but often enter aquatic environments via soil percolation, air drift, and surface runoff, affecting abundance and diversity of nontarget species, producing complex effects on the ecosystems, and altering trophic interactions (Rand et al. 1995). In addition, many pesticides eventually end up in groundwater, where their residues or transformation products may remain for years (Belfroid et al. 1998).

Ours was the first systematic data collection by direct measurement of insecticide residues in the river over the course of a 2-yr monitoring program with monthly samplings in S1, S4, and S5 (Rovedatti et al. 2001). Of the samples analyzed, 35% had organochlorine pesticides in a concentration greater than 0.1 $\mu g\,L^{-1}$ (ppb). Organophosphates were not found, possibly due to their low persistence because of their short half-lives in aquatic environments.

At all locations, pesticide levels varied from 40 to 400 times above the MPQ for protection of freshwater life according to current Argentine legislation. Throughout the study period, DDT and its metabolite, DDE, were found only in S1 and gamma-chlordane in S4; heptachlor was found in 50% of the samples of S4 and in 35% of S5, while HCH isomers were detected in 38% of S4 and 45% of S5 samples. High levels of some pesticides were simultaneously detected but with greater frequency downstream (S4 and S5). Neither temporal nor spatial trends were found. There was no relationship between the detected pesticide residues and the application season for farming purposes.

These data are in agreement with those that result from the remaining physicochemical information, confirming that the quality of the river water is considerably deteriorated between S4 and S5, providing additional impact on the biota and bringing a potential risk to human health.

It must be mentioned that, since the 1970s, the import, manufacture, commercialization, and agricultural use of products based on DDT, heptachlor, chlordane, dieldrin, aldrin, and HCH (except its pure gamma isomer) was forbidden in Argentina (Piazza et al. 2000). In this respect, it is interesting to mention that García Fernández et al. (1979), after an extensive screening of major organochlorine pesticides in two main large rivers of the same basin, have found high levels of α-, β-, and γ-HCH [up to 0.86 $\mu g\,L^{-1}$ (ppb)], heptachlor and its epoxide [up to 1.12 $\mu g\,L^{-1}$ (ppb)], and pp′-DDT [up to 5.6 $\mu g\,L^{-1}$ (ppb)].

Organochlorine pesticide levels exceeding guidelines were also detected in coastal water samples of Río de la Plata, the final receptor of the Reconquista River water (AA-AGOSBA et al. 1997; AGOSBA et al. 1992; Colombo et al. 1990). However, in the principal watercourse of that river, those values remain below the regulatory limits as a result of dilution processes (Comisión Administradora del Río de la Plata 1989).

The natural and anthropogenic pollution of groundwater in Argentina by heavy metals and pesticides has been recently reported (Ares et al. 1999; Farías et al. 2003; Loewy et al. 1999; Marteau et al. 1999; Momo et al. 1999). However, the kinetics of the interactions between pollutants in groundwater and surface water for this particular system was not studied.

Our previous data and the present results indicate that the river receives intermittent contaminant inputs of pesticides. Although the sampling sites located after the Morón stream's discharges appeared as the more polluted, upstream points proved to be also polluted, although less extensively than in the remaining locations. It must be pointed out that our research was restricted to only two groups of insecticides. The overall picture will change if the last-generation pesticides (e.g., pyrethroids), currently being used, is included for consideration.

Turbidity This variable is a relevant factor both in primary productivity and in the transport and bioavailability of several chemicals such as heavy metals. Tubidity results from the scattering and absorption of incident light by particulate matter originated from erosion in the watershed and/or resuspension of particles deposited in the riverbed.

Turbidity was one of the parameters that remained almost constant, without important spatial or temporal changes within each sampling series. This finding supports the conclusion that the detected differences in algal biomass (see following) may be attributed to the effects of either xenobiotics, seasonal temperature, or nutrient concentration variations rather than to differential penetration of light.

The amounts of total dissolved solids were variable, with a tendency to increase in the sampling stations located close to the mouth of the river, but with a very irregular spatial profile, closely dependent on rainfall. The mean values oscillated between 250 and 1100 $mg\,L^{-1}$. The suspended matter was a minor proportion of the total solids and showed a tendency to diminish in S4 and S5.

Water Quality Chemical Indexes Several indexes have been proposed and applied as relative indicators of water quality. They have the advantage of summarizing large sets of data into simple numeric expressions.

An ecological classification of riverine systems based on the concentrations of dissolved heavy metals has been proposed by Wachs (1998). According to that classification, a water body *excessively polluted* (grade IV) shows the following concentration profile (in $\mu g\,L^{-1}$): Cd > 2, Cr > 8, Cu > 20, Pb > 15, and Zn > 140. In the case of the Reconquista River, those limits were exceeded in most cases by 32%–47%.

The ICA (*Index of Water Pollution*) (Berón 1984) is an index of organic pollution. Determined by temperature, Cl^-, $N\text{-}NH_4^+$, BOD, and DO, it varies between 0 and 10 (worst–best condition); an index of 10 corresponds to a sample with a pollution status equivalent to a sewer effluent. The ICA was 6.1–8.9 in Cascallares (S1), 3–6 in Paso del Rey (S2), 3–5 in S3, 1.7–3.8 in San Martín (S4), and 1.2–4.4 in Bancalari (S5) (Olguín et al. 1999; Topalián and Castañé 2003). By means of the same index, other authors have shown that the quality of the water has declined since our samplings (Arreghini et al. 1997).

The ICAPI (Lacoste and Collasius 1995) (*Index of Water Pollution by Industries*) is another index used to quantify industrial contamination; it relies on the concentration of phenols, detergents, and heavy metals (As, Cd, Cr, Cu, Hg, Pb, Zn), DO, and COD. The values of the ICAPI also vary within a 0–10 (worst–best) scale with 10 equivalent to an original purity state and 0 corresponding to an untreated industrial effluent. The ICAPI was 3.4–5.0 in Cascallares (S1), 0.4–2.0 in San Martín (S4), and 1–2 in Bancalari (S5) (Olguín et al. 1999; Topalián and Castañé 2003).

When the ICA indexes were grouped by season, it was found that in most cases the worst condition corresponded to the autumn–winter period. The ICAPI index did not show similar fluctuations, suggesting that the industrial effluents were spilled regularly into the river all year long (Topalián and Castañé 2003).

B. Biotic Parameters

Phytoplankton In spite of the amount and range of domestic, agricultural, and industrial pollutants poured into the river, considerably diverse algal communities were found, even at points such as Bancalari (S5) where the pollution indexes showed elevated deterioration of water quality (in particular, permanent anoxic condition and extremely high concentrations of heavy metals).

A maximum of 160 algal species distributed among 92 genera were recorded in different points of the river (Castañé et al. 1998b; Loez 1995). The number of genera was similar or even lower than those reported for other rivers of the same basin located in areas of comparable geological and climatic characteristics, studied within the same time (Conforti et al. 1995; del Giorgio et al. 1991), and receiving comparable types of discharges to those poured into the Reconquista River. The differences described in algal diversity may be interpreted as a consequence of the higher quantities of pollutants in our river.

The total phytoplankton density fluctuated extremely both in time and in space, being usually in the range of 1,600–25,000 cells mL^{-1} but reaching extreme values of 13×10^8 cells mL^{-1}, obviously the highest end of that range was registered when blooms occurred. The general trend was in the direction of an increase between S1 and S4 coincident with the augmentation of the nutrient concentrations, while it declined in the vicinity of the mouth (S5) as a response to high pollution stress, probably due to the synergistic effect of elevated hardness and heavy metals concentration. Similarly, the specific richness and diversity were also reduced; density showed climatic oscillations, the highest values being found in spring.

The biomass of Bacillariophyceae and Chlorophyceae (associated with organic pollution) was in general dominant, but the density of Cyanophyceae and Euglenophyceae (indicators of abundant organic matter and low DO) increased in urban and industrialized zones of the river's basin (Loez and Topalián 1999).

In summer and winter, frequent blooms of some species in limited areas of the river were recorded, associated with an elevation of chlorophyll *a* levels. However, it is interesting to mention that under "normal" conditions the algal biomass did not show significant changes, remaining approximately stable.

Although total concentrations of nutrients were always rather high, levels of chlorophyll *a* generally remained moderate because of high discharge rates and large amounts of nonliving suspended solids, causing reduced growth of algae by washout and light limitation. O'Farrell et al. (2002) have observed this behavior in the Luján River.

It is worth mentioning that after the output of the Morón stream, several species of *Pediastrum* showed important morphological anomalies (Loez 1995).

Only a very small proportion of the species were restricted to a particular site, while the majority were found in all sampling points, although, as already stated, with differences in their densities. Throughout our study most of the species found were typical cosmopolitans in eutrophic rivers. From the density values of the different algal groups, the degree of eutrophication of each site was estimated by means of the Nygaard Index. It resulted in 11.3, which was evidence of an eutrophic environment with organic pollution.

The simplest interpretation of these findings relates the increase in species diversity found in S1–S4 to the gradual nutrient enrichment of the water, especially by N- and P-containing compounds in the water or, alternatively, as an evidence of adaptation of algal communities to high pollutant levels, thus explaining the fact that certain assemblages of species were registered regularly in all seasons, in all sampling stations and during the entire monitoring program. That was the case of *Aulacoseira granulata* var. *angustissima, Chlorella vulgaris, Crucigenia crucifera, Cyclotella meneghiniana, Euglena acus, Microcystis aeruginosa, Monoraphidium arcuatum, Nitzschia palea*, and *Oscillatoria chlorina,* which occurred regularly. These species can be considered indicators of pollution or tolerant to the presence of pollutants.

Relatively abundant and diverse algal communities were found even in Bancalari (S5), a point characterized by its permanent anoxic condition and high dissolved Zn^{2+} concentration, 1200–4000 $\mu g\ L^{-1}$ (ppb) (Loez 1995; Loez and Salibián 1990; Loez et al. 1995, 1998). In this respect it could be concluded that metals were not completely available and/or that some species have adapted to them through an increase in their toxicity thresholds. Some of the results obtained from the bioassays (see below) strongly suggest that possibility.

It is interesting to mention the report by Vigna et al. (2002) of the isolation from the water of the Reconquista River of a coccoid algal species that showed the capacity to degrade gas oil as its unique source of energy and carbon.

Zooplankton The mean total abundance of zooplankton was 70–120 individuals L^{-1} with rotifers as dominant species. Kuczynski (1991b) reported the same trend in samplings carried out 10 years ago. Among crustaceans, larval copepods (nauplii) and cyclopodean copepods were found though in higher density than cladocerans and calanoid copepods.

The annual mean density of most of the zooplankton groups showed a clear tendency to decrease from S1 to S5, mainly due to a reduction in the cladocerans (Castañé et al. 1998b). The same trend was reported by Kuczynski (1984) in samplings carried out in 1982.

The analysis of the changes in the density of the zooplankton communities has shown significantly positive correlations with DO and negative correlations with pollution-associated parameters (N-NH_4^+, hardness, phosphates, COD, BOD, low ICA) (Olguín et al. 1999; Puig 1997a; Puig and Salibián 1999).

This profile is not homologous to that described for the phytoplankton. In this case, a density decrease was recorded and yet their available food in terms of phytoplankton density and algal biomass did not reveal any important changes. Therefore, it is reasonable to suggest that the interpretation of the changes in the zooplankton profile would rather be caused by the deterioration of the chemical quality of the water, specially due to increased hardness coupled to changes in the concentrations of different substances (e.g., oxygen, N-NH_4^+) up to toxic levels (Puig 1997b).

As in the case of the phytoplankton, the spatial changes in the pattern of some groups of zooplankton (for instance, dominance of rotifers by crustacean decrease in Bancalari) might be interpreted as a consequence of specific adaptations to particular physicochemical environments.

Thus, the rotifer–crustaceans relationship could be considered as an indicator of water quality, increasing in the most deteriorated sites, and characterized by high organic and industrial pollution and by conditions of permanent anoxia, such as in S5 (Olguín-Salinas et al. 1999).

Microbiology The water quality of urban rivers in Argentina is worsening, not only because of increased input of chemicals but also because of the large amounts of discharges from septic tanks, raw sewage, and, in much less extent, from sewage treatment plants. From the public health standpoint we considered an important objective was to examine whether the Reconquista River water is polluted with feces. This is the first report referring to a long-term monitoring of bacterial pollution in the surface water samples of the Reconquista River collected monthly in three critical sites: Cascallares (S1), San Martín (S4), and Bancalari (S5). Preliminary reports of the results were published elsewhere. (Martínez and Salibián 1995a,b).

Among the indicators currently used to estimate fecal pollution of water are N-NH_4^+ and the concentration of coliforms, total and fecal, calculated from the most probable number (MPN). The results (Table 2) were alarming. In general, S1 was a relatively less polluted point compared with S4–S5. The bacteriological diagnosis of samples has shown the dominant presence of *Escherichia*

Table 2. Most probable number index (MPN) (100 mL^{-1}), $N\text{-}NH_4^+$ (mg N L^{-1}), BOD and COD (mg O_2 L^{-1}) in water samples of the Reconquista River taken from Cascallares (S1), San Martín (S4) and Bancalari (S5).

Month	Parameter	S1	S4	S5
October	MPN	1.70×10^4	2.20×10^6	8.00×10^6
	$N\text{-}NH_4^+$	1.70	15.80	15.20
	BOD	7.80	61.00	66.00
	COD	103.00	300.00	250.00
November	MPN	9.00×10^2	2.20×10^6	1.70×10^6
	$N\text{-}NH_4^+$	0.60	15.60	19.60
	BOD	5.80	55.70	43.50
	COD	79.00	237.00	225.00
December	MPN	3.00×10^4	3.30×10^7	3.00×10^6
	$N\text{-}NH_4^+$	1.20	11.10	6.25
	BOD	3.90	65.50	42.50
	COD	46.00	351.00	339.00
February	MPN	3.50×10^3	7.00×10^6	8.00×10^6
	$N\text{-}NH_4^+$	0.35	20.00	20.10
	BOD	10.80	104.50	85.50
	COD	70.00	310.00	324.00
March	MPN	1.1×10^4	5.0×10^6	5.0×10^6
	$N\text{-}NH_4^+$	0.90	2.90	2.80
	BOD	13.00	59.80	36.00
	COD	71.00	213.00	209.00
April	MPN	2.20×10^4	9.00×10^5	7.00×10^5
	$N\text{-}NH_4^+$	0.00	0.90	2.20
	BOD	3.50	10.50	9.00
	COD	55.00	46.00	59.00
May	MPN	2.20×10^4	5.00×10^6	3.00×10^6
	$N\text{-}NH_4^+$	0.50	4.90	4.80
	BOD	5.70	26.00	39.50
	COD	55.00	190.00	228.00
June	MPN	5.00×10^3	1.60×10^7	1.10×10^7
	$N\text{-}NH_4^+$	0.60	9.30	10.00
	BOD	0.60	15.50	49.00
	COD	45.00	99.00	198.00
August	MPN	5.00×10^3	3.00×10^5	1.10×10^7
	$N\text{-}NH_4^+$	0.20	15.80	15.40
	BOD	3.00	58.00	65.50
	COD	59.00	303.00	312.00
September	MPN	2.80×10^3	1.10×10^7	9.00×10^6
	$N\text{-}NH_4^+$	0.50	17.50	19.00
	BOD	0.90	81.00	75.00
	COD	77.00	354.00	321.00

Table 2. (Continued)

Month	Parameter	S1	S4	S5
October	MPN	1.70×10^4	3.00×10^6	2.20×10^6
	$N\text{-}NH_4^+$	0.50	12.30	14.10
	BOD	4.20	51.00	52.00
	COD	50.00	165.00	150.00
November	MPN	2.20×10^4	1.40×10^6	5.00×10^6
	$N\text{-}NH_4^+$	0.3	9.3	10.1
	BOD	n.m.	n.m.	n.m.
	COD	n.m.	n.m.	n.m.
December	MPN	3.00×10^4	2.20×10^6	9.00×10^6
	$N\text{-}NH_4^+$	n.m.	n.m.	n.m.
	BOD	n.m.	n.m.	n.m.
	COD	n.m.	n.m.	n.m.

BOD: biological oxygen demand; COD: chemical oxygen demand; DO: dissolved oxygen; n.m., not measured.

coli in all samples, together, but in lower proportion, with *Enterobacter* sp., *Klebsiella* sp., *Salmonella* sp., and *Shigella* sp.

The pollution in terms of fecal bacteria increased downstream, especially after the output of the Morón stream into the river. The results also indicated a reasonable correlation of the bacterial pollution with $N\text{-}NH_4^+$ concentration, BOD, and COD values of the samples. All MPNs were well above the local and international water quality guidelines. The results show that the river presents (a) a heavy bacteriological pollution level at all studied sites, reflected by an elevated number of fecal coliforms and (b) important fluctuations of the pollution both spatially and temporally.

Finally, it is interesting to mention that the physicochemical profile of the studied samples showed elevated concentrations of heavy metals, which suggests an adaptation of bacterial strains to particular environments. Vidal et al. (1993) have isolated, from the water of the river, *Pseudomonas* and *Alcaligenes*, which are characterized by their resistance to high levels of Cr, Hg, Cd, and Pb.

Toxicity Bioassays It is well known that bioassays with sensitive species representative of different organization levels as sentinel organisms are useful tools for the quantitative evaluation of the toxicity of complex mixtures. In acute assays the usual endpoint is the survival rate after 96 hr exposure; in longer assays, under sublethal conditions, biomarkers have the potential to be integrative indicators of the suborganismal effects of the simultaneous impacts of multiple xenobiotics and/or environmental stress factors on the test organism as a whole. Changes in different parameters in the sentinel organism (biochemical, physiological, morphological, behavioral, *etc.*) may be appropriate bioindicators of toxic effects, exposure, or susceptibility to toxic substances; and they also may serve to identify reliable sites that represent a threat to the health and survival of contaminant-exposed populations (Porta 1996; WHO 1993).

For acute toxicity bioassays of the Reconquista River water samples (S1, S4, and S5), algal and animal species as test organisms were used applying protocols and techniques developed in our laboratory (Ferrari et al. 1997, 1998; Olguín et al. 2000). In both algal and animal assays, a standard reference toxicant was included in the test batteries at concentrations close to the 96-hr LC_{50} or IC_{50} values for each species.

Bioassays with Algae Because of their sensitivity, algae are primary producers particularly useful as test organisms in toxicity bioassays in both unispecific and plurispecific systems. They allow quick responses evaluated by means of different endpoints (stimulation or inhibition of the growth rate, changes in biomass and diversity, or alterations in the community structure). In addition, algae are sensitive indicators of the combined effect of changes in physical parameters and nutrient concentration and of inputs of toxicants.

Two types of algal assays were performed. One group of prolonged assays determined the impact of Zn on the native seasonally taken samples of whole phytoplanktonic communities. These samples were incubated in the laboratory in the presence of different concentrations of the metal for 1 month. Responses of these multispecific bioassays were variable and concentration dependent, having observed specific stimulations or inhibitions of the growth rate of certain species, thus allowing the determination, for each assayed concentration of the metal, of particular Zn-tolerant associations characterized by typical diversity, richness, and evenness. Chlorophyceae and Bacillariophyceae were dominant algal classes, with Diatomophyceae the most toxicant-sensitive algae.

These results may help explain the fact that in spite of the high concentrations of some heavy metals found in heavily polluted points of the river, such as Bancalari (S5), there was a considerably diverse algal community (Loez et al. 1995, 1998; Topalián and Loez 1995).

The second type of algal assays were performed with three Chlorophyceae, *Chlorella vulgaris*, *C. pyrenoidosa*, and *Scenedesmus acutus*, representative species of the phytoplankton communities present in almost all sampling sites of the river. Algae used in these unispecific toxicity assays were previously cultivated in axenic conditions and were free of previous contact with xenobiotics. Assays with pure media and with samples of different points of the river were run simultaneously. Cultures showed in all cases stimulation of the growth rate, which was attributed to the presence of high quantities of nutrients, mainly phosphates and N-NH_4^+, showing the greatest increase in S4 samples whereas those from S5 showed the least increase. These results corresponded with the chemical composition of the samples and thus explained the simultaneous presence of important quantities of toxicants. *S. acutus* was the most sensitive species (Ferrari et al. 1998; Olguín et al. 2000; Olguin-Salinas et al. 1999).

Bioassays with Animals Among consumers, we used two teleost species (fry and juveniles of *Cnesterodon decemmaculatus* and juveniles of *Cyprinus carpio*) (de la Torre 2001; de la Torre et al. 1997, 2000a; Ferrari et al. 1998; García et al. 1998). *C. decemmaculatus* is a small ovoviviparous fish of the indigenous fauna present in all aquatic environments of the pampasic plains that is also frequently found in low-level polluted sites of the river. The biology of this species is very well known. The common carp *C. carpio* is an introduced species, standardized by international agencies as a sentinel organism for aquatic toxicity tests. All the animals used in laboratory tests were also specimens reared in controlled conditions, free from previous contact with xenobiotics.

Acute and chronic sublethal toxicity fish bioassays were performed following two protocols: (a) bioassays in laboratory conditions (larvae and juveniles) and (b) *in situ* caged adult fish. The cages were submerged in the water at different sampling stations along the river.

In chronic sublethal assays, several selected biochemical and physiological parameters as biomarkers of the function of several systems of the animals and somatic indexes (liver somatic index, condition factor) were measured. All biomarkers and indexes were measured in the same animals (de la Torre 2001; de la Torre et al. 1999, 2000b).

Premetamorphic larvae of a native toad (*Bufo arenarum*) (de la Torre 2001; Ferrari et al. 1998) were also used. This animal is a species widely distributed in the region and characterized by its fully aquatic early development. We have shown that the young tadpoles of these animals are useful biological tools for performing standardized acute toxicity assessment tests (Demichelis et al. 2001; Salibián 1999); their sensitivity was greater than that shown in young fish.

Both fish and tadpole assays enabled us to discriminate between highly polluted sites of the river (S4–S5) and those whose degree of deterioration was not so remarkable (S1) (Figures 3, 4). The assays made possible the detection of temporal variations in water pollution at a given particular site.

Chronic bioassays with fish had three phases. The first was the acclimation phase, followed by the exposure and recovery periods, the last one was designed to evaluate the reversibility of the altered biomarkers by exposing the test animals to diluted samples or to clean water. All assays included a number of appropriate controls. In these assays, gill ATPases and liver aminotransferases were the most affected variables pointing to the impairing effect of the river water on the respiratory, ionoregulatory, and excretory capacity of the gills as well as the liver function involved in protein and carbohydrate metabolism and in xenobiotic conversions. The magnitude of brain acetylcholinesterase inhibition was a valuable marker for heavy metals as well as of previous contact with organophosphorous and neurotoxic contaminants (de la Torre et al. 2002); the activity of that enzyme is vital to normal behavior and muscular function.

It is important to point out that biomarkers were altered in all cases, even in those assays performed with samples from sites considered to have low contamination (e.g., Cascallares). Nevertheless, most of the detected alterations were reversible at the third phase of the assays when river water samples were diluted or replaced with clean media.

During the assays, samples of water were always taken simultaneously, so that we were able to correlate by means of a stepwise analysis the response of the animals with the relevant components of the physicochemical profile of each

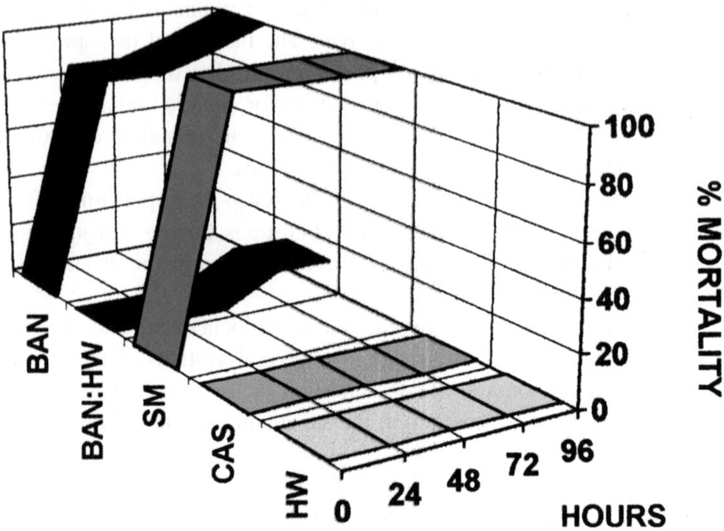

Fig. 3. Cumulative mortality curves of premetamorphic *Bufo arenarum* tadpoles held 96 hr in control media (HW) and in samples of the Reconquista River. BAN (S5); BAN:HW (S5 diluted 1 + 1 with hard water) SM (S4) CAS (S1) HW (EPA hard water).
Source: Demichelis et al. (2001)

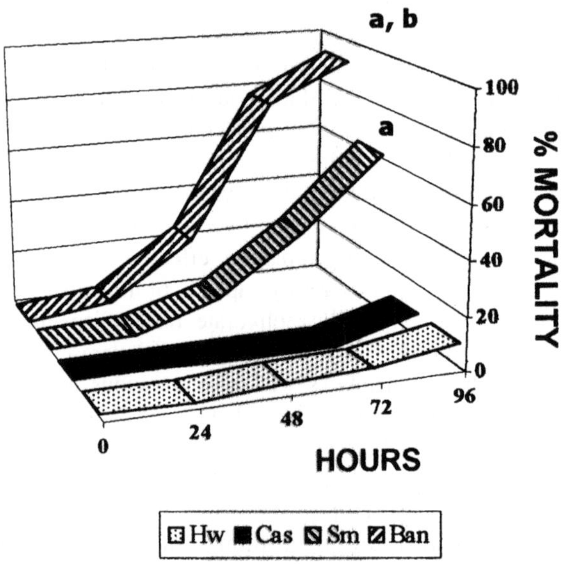

Fig. 4. Cumulative mortality of juvenile *Cnesterodon decemmaculatus* exposed to samples of Reconquista River water. Hw, EPA hard water; Cas, Cascallares (S1); Sm, San Martín (S4); Ban, Bancalari (S5); a, significantly different from Hw and Cas; b, significantly different from Sm.
Source: de la Torre et al. (1997)

sample. This, in turn, allowed determining predictive regression equations corresponding to mortality and to a particular set of physicochemical parameters of the tested water sample.

There was variability in the toxicity of water depending on the physicochemical profile of the assayed samples, which was considered as a relevant result of our work, showing the importance of making precise diagnosis after an integrative protocol of physicochemical parameters with the responses of intact aquatic organisms. In our case, results of the models constructed after the bioassays with animal species showed that acute toxicity of the river water was related to an integrated response to pH, $N\text{-}NO_2^-$, $N\text{-}NH_4^+$, Cl^-, BOD, and heavy metals; in other words, both the domestic and industrial effluents were responsible for the acute toxicity of water.

Finally, it must be noted that our results showed the usefulness of indigenous fish and amphibian species as reliable test organisms for environmental biomonitoring, equally sensitive to toxic mixtures as are species of introduced biota.

VI. Conclusions

The Reconquista River is a watercourse that shows an important and sustained chemical and bacteriological stress, resulting in a critical condition. No part of the river can be considered pollutant free. The diverse and complex mixtures of untreated effluents that are discharged into the river, the highly polluted tributaries, and its reduced flow have caused it to exceed its dilution and self-purification capacity.

An earlier study (Loez and Salibián 1990) showed that spatially the river may be considered as a succession of several rivers, each one with its own particular chemical and biological characteristics. One of them can be the S1–S2 area; the second "river" is doubtless S4–S5, whereas S3 may be considered as the portion of the river in transition that links the less-polluted "river" to the highly polluted area. The site of the Morón stream entrance into the main course of the river makes a clear-cut separation between S1–S2 and S4–S5 areas. This conclusion was confirmed in all subsequent studies carried out in our laboratory and by other investigators.

The results presented here confirm that in its present condition the Reconquista River may not be used for any human purpose and that it shows serious risks for the preservation of biodiversity.

The river was considered by several authors as a "dead river." Notwithstanding, our conclusion is that its water quality may be improved if adequate remediation measures are taken. The fact that a large amount of information is now available facilitates the design and monitoring of the efectiveness of those measures.

In spite of the contamination, the behavior of the plankton communities indicates their adaptation to survive in very adverse conditions. In this respect, there is no doubt that this group of microorganisms will be fundamental in any recuperation process of water quality.

When considering physicochemical and biological parameters altogether in a spatial and temporal perspective, it is possible to reach two main conclusions: first, that there was a progressive but sustained alteration of the water quality downriver, especially after the confluence of the Morón stream, and second, that the deterioration has increased significantly during the past 15 years. More recently, other authors (de Cabo et al. 2000), after a shorter monitoring, contemporary to ours, reported results that confirm our conclusions.

Summary

The situation of most of the Argentine rivers is very serious due to the amount and range of pollutants, principally as a consequence of industrial development with an inadequate regulatory framework and a deficit of decades in matters of sanitary substructure and waste treatment.

Freshwater quality monitoring in Argentina was based on water chemistry and bacteriology, with measurements of only the main variables required for the determination of quality indexes. A multidisciplinary approach considering simultaneous evaluation of a number of factors and processes that in an integrative picture determine its characteristics was poorly developed or lacking. The use of biota for monitoring the aquatic environments has been relatively uncommon compared to abiotic variables.

The Reconquista River is a typical lowland watercourse situated in the Buenos Aires Province. Located in a temperate subtropical region, it flows into an international river, Río de la Plata, which is part of the second largest hydrogeographic system of South America, after the Amazon, and the fifth largest in the world. The river receives the output of 80 small tributaries, and one of them, the Morón creek, should be highlighted as it marks the limit between the medium and the lower sections of the river. During the dry season the entire stream is composed of sewage and industrial wastes and known as an "open sewer."

The river is the second most polluted waterway of Argentina. There is a great variety of industrial activities settled on its basin. Some 10,000 plants, most of them located on the margins of the river, discharge their untreated effluents into the river and use large quantities of water in processing, cooling, and cleaning. Approximately 20% of these industries discharge a total BOD load of approximately 150,000 kg d^{-1}, which is equivalent to an organic loading capacity of 2.5 million population.

To monitor water quality of the main course of the Reconquista River, the following principal approaches have been adopted: (a) measurement of approximately 30 physical and chemical variables, (b) determination of biological parameters (phytoplankton and zooplankton abundance, diversity, and community structure, microbiology), (c) acute and prolonged *in situ* and laboratory toxicity tests with algae, tadpoles, and fish (fry and juveniles) as sentinel organisms, (d) monitoring of alterations in specific physiological and biochemical markers of exposure in fish, and (e) determination of water quality indexes based on the physicochemical profile of the samples.

The studies were carried out during a span of 15 years on samples regularly taken in three to five sites covering the length of the river's main bed. The most relevant outcomes of the study can be summarized as follows.

Spatial variation of the DO was from 7–8 mg L^{-1} in S1 down to 0–0.3 mg L^{-1} in S5, indicating water in Bancalari was in permanent anoxia.

COD/BOD ratios oscillated between 11 (in S1) and 4 (in S4–S5), suggesting the presence of important amounts of nonbiodegradable organic matter.

The presence of domestic sewage indicators and municipal wastes as chlorides, orthophosphates, inorganic N compounds (NH_4^+, NO_2^-, NO_3^-) and phenols were found in all samples, with a clear-cut increase in concentration down river up to values well above MPQs.

Total heavy metal concentrations always exceeded widely MPQs established by Argentine law for protection of freshwater life.

At all locations organochlorine insecticide levels varied between 40- and 400 fold above legal limits.

When considering physicochemical parameters altogether in a spatial and temporal perspective, it becomes evident that there was a progressive but sustained alteration of the water quality downriver, especially after the confluence of the Morón creek, and, that the deterioration increased with time.

River water showed an alarming degree of bacterial pollution, which was temporally and spatially highly variable.

Environmental physicochemical toxicity data were closely correlated with the results of both acute and sublethal chronic toxicity bioassays.

It was concluded that only an integrated analysis of chemical and biological parameters coupled with toxicity tests will offer a realistic view of the state of water quality in the Reconquista River.

Acknowledgments

The results reported and discussed in this review were generated in a number of projects supported by grants to the author from the National Research Council of Scientific and Technological Research (CONICET, Argentina), the Scientific Research Commission of the Buenos Aires Province (CIC-Buenos Aires, Argentina), and the Basic Sciences Department (Programa de Incentivos) of the National University of Luján, Argentina. The author thanks all colleagues that participated in the studies whose results have been reported in this article. Their names are included as coauthors of the papers listed below. The very efficient and enthusiastic technical assistance of Roberto Yoshihara and Javier Katz is also acknowledged. The author thanks Ms. Cecilia Moreno of the translators' team of the CIC-Buenos Aires, for her assistance in the preparation of the English version of the manuscript.

References

AA-AGOSBA-OSN-SIHN (República Argentina) (1997) Río de la Plata. Calidad de las aguas de la franja costera Sur del río de la Plata (San Fernando-Magdalena). Buenos Aires.

AGOSBA-OSN-SIHN (República Argentina) (1992) Río de la Plata. Calidad de las aguas. Franja costera Sur (San Isidro-Magdalena). Buenos Aires.

Alsina MG, Herrero AC (2000) Relevamiento de industrias en la cuenca del río Reconquista y georreferenciación de las de tercera Categoría según el nivel de complejidad ambiental (Ley 11459). Relación entre actividad industrial y grado de contaminación de la cuenca. Actas Simposio Latinoamer Percepción Remota (Iguazú, Argentina): 867–878.

Ares JO, Miglierina AM, Sánchez R (1999) Patterns of groundwater concentration and fate of lindane in an irrigated semiarid area in Argentina. Environ Toxicol Chem 18:1354–1361.

Arreghini S, de Cabo L, Iorio AF de, Rendina A, Bargiela M, Godoy di Pace G, Corujeira A, Vella R, Bonetto C (1997) Evaluación de la calidad de aguas del río Reconquista a través de índices. Resúm Congr Internac sobre Aguas. Facultad de Derecho y Ciencias Sociales, Universidad de Buenos Aires, III. 24.

Belfroid AC, van Drunen M, Beek MA, van Gestel, CM, van Hattum B (1998) Relative risks of transformation products of pesticides for aquatic ecosystems. Sci Total Environ 222:167–183.

Berón L (1984) Evaluación de la calidad de las aguas de los ríos de la Plata y Matanza-Riachuelo mediante la utilización de índices de calidad de agua. Secretaría de Vivienda y Ordenamiento Ambiental. Ministerio de Salud y Acción Social, Argentina.

Bilos C, Colombo JC, Presa MJR (1998) Trace metals in suspended particles, sediments and asiatic clams (*Corbicula fluminea*) of the Río de la Plata estuary. Environ Pollut 99:1–11.

Borthagaray JM, Fernández Prini R, Igarzábal de Nistal M, San Román E, Tudino M (eds) (2001) Diagnóstico Ambiental del Area Metropolitana de Buenos Aires. FADU-UBA, Buenos Aires.

Brailovsky AE, Foguelman D (1992) Memoria verde. Sudamericana, Buenos Aires.

Castañé PM, Topalián ML, Rovedatti MG, Salibián A (1998a) Impact of human activities on the water quality of the Reconquista River (Buenos Aires, Argentina). Verh Int Verein Limnol 26:1206–1208.

Castañé PM, Loez CR, Olguín HF, Puig A, Rovedatti MG, Topalián ML, Salibián A (1998b) Caracterización y variación espacial de parámetros fisicoquímicos y del plancton en un río urbano contaminado (Río Reconquista, Argentina) Rev Int Contam Ambient 14:69–77.

Catoggio JA (1990) Contaminación del agua. Causas de la contaminación de aguas superficiales y subterráneas. Precipitaciones ácidas. Eutroficación; polución costera. In: Instituto de Estudios e Investigaciones sobre el Medio Ambiente (ed), Latinoamérica. Medio Ambiente y Desarrollo, pp 137–155.

Colombo JC, Pelletier E, Brochu C, Khalil MF (1989) Determination of hydrocarbon sources using *n*-alkane and polyaromatic hydrocarbon distribution indexes. Case study: Rio de la Plata estuary, Argentina. Environ Sci Technol 23:888–894.

Colombo JC, Khalil MF, Arnac M, Horth AC (1990) Distribution of chlorinated pesticides and individual polychlorinated biphenyls in biotic and abiotic compartments of the Rio de La Plata, Argentina. Environ Sci Technol 24:498–505.

Colombo JC, Brochu C, Bilos C, Landoni P, Moore S (1997) Long-term accumulation of individual PCBs, dioxins, furans, and trace metals in Asiatic clams from the Rio de la Plata estuary, Argentina. Environ Sci Technol 31:3551–3557.

Comisión Administradora del Río de la Plata (1989) Estudio para la evaluación de la contaminación en el río de la Plata. Informe de avance, vols I and II. Buenos Aires.
Conforti V, Alberghina J, González Urda E (1995) Structural changes and dynamics of the phytoplankton along a highly polluted lowland river of Argentina. J Aquat Ecosyst Health 4:59–75.
de Cabo L, Arreghini S, Iorio AF de, Rendina A, Bargiela M, Vella R, Bonetto C (2000) Impact of the Morón stream on water quality of the Reconquista River (Buenos Aires, Argentina). Rev Mus Argentino Cs Nat n s 2:123–130.
de la Torre FR (2001) Estudio integrado de la contaminación acuática mediante bioensayos y parámetros fisiológicos y bioquímicos indicadores de estrés ambiental. Doctoral dissertation, Faculty of Exact and Natural Sciences, University of Buenos Aires.
de la Torre FR, Salibián A, Ferrari L (1999) Enzyme activities as biomarkers of freshwater pollution: responses of fish branchial (Na + K)-ATPase and liver transaminases. Environ Toxicol 14:313–319.
de la Torre FR, Demichelis SO, Ferrari L, Salibián A (1997) Toxicity of Reconquista river water: bioassays with juvenile *Cnesterodon decemmaculatus*. Bull Environ Contam Toxicol 58:558–565.
de la Torre FR, Ferrari L, Salibián A (2000a) Long-term *in situ* water toxicity bioassays in the Reconquista river (Argentina) with *Cyprinus carpio* as sentinel organism. Water Air Soil Pollut 121:205–215.
de la Torre FR, Salibián A, Ferrari L (2000b) Biomarkers assessment in juvenile *Cyprinus carpio* exposed to waterborne cadmium. Environ Pollut 109:277–282.
de la Torre FR, Ferrari L, Salibián A (2002) Freshwater pollution biomarker: response of brain acetylcholinesterase activity in two fish species. Comp Biochem Physiol 131C: 271–280.
del Giorgio PA, Vinocur AL, Lombardo R, Tell HG (1991) Progressive changes in the structure and dynamics of the phytoplankton community along a pollution gradient in a lowland river. A multivariate approach. Hydrobiologia 224:129–154.
Demichelis SO, de la Torre FR, Ferrari L, García ME, Salibián A (2001) The tadpole assay: its application to water toxicity assessment of a polluted urban river. Environ Monit Assess 68:63–73.
Escalona L, Winchester JW (1994) La tendencia de la urbanización en Sur América: inferencias a partir de una base de datos del "World Resources Institute." Interciencia 19:64–71.
Faggi AM, Arriaga MO, Aliscioni SS (1999) Composición florística de las riberas del río Reconquista y sus alteraciones antrópicas. Rev Mus Argentino Cs Nat ns 1(1):1–6.
Farías SS, Casa VA, Vázquez C, Ferpozzi L, Pucci GN, Cohen IM (2003) Natural contamination with arsenic and other trace elements in ground waters of Argentina Pampean Plain. Sci Total Environ 309:187–199.
Fernández Cirelli A (Comp) (1998). Agua. Problemática regional. Enfoques y perspectivas en el aprovechamiento de recursos hídricos. EUDEBA, Buenos Aires.
Ferrari L, Demichelis SO, García ME, de la Torre FR, Salibián A (1997) Premetamorphic anuran tadpoles as test organism for an acute aquatic toxicity test. Environ Toxicol Water Qual 12:118–121.
Ferrari L, García ME, de la Torre FR, Demichelis SO (1998) Evaluación ecotoxicológica del agua en un río urbano mediante bioensayos con especies nativas. Rev Mus Argentino Cs Nat, Volumen Extra (Ecotoxicología). Nueva Serie 148:1–16.
Finkelman J (1996) Chemical safety and health in Latin America: an overview. Sci Total Environ 188(suppl 1):S3–S29.

Foguelman D, González Urda E (1994) El agua en Argentina. Prociencia CONICET, Buenos Aires.

García ME, Demichelis SO, de la Torre FR, Ferrari L (1998) Freshwater toxicity to *Cnesterodon* sp.; bioassays with water from the Reconquista river. Verh Verein Limnol 26:1216–1218.

García Fernández JC, Marzi A, Casabella A, Roses O, Guatelli M, Villaamil E (1979) Pesticidas organoclorados en el agua de los ríos Paraná y Uruguay. Ecotoxicología 1:51–78.

Hair JF, Anderson RE, Tatham RL, Black WC (1995) Multivariate data analysis. Prentice Hall, Englewood Cliffs, NJ.

Hajek ER (Comp) (1995) Pobreza y medio ambiente en América Latina. K Adenauer-Stiftung-CIEDLA, Buenos Aires.

INDEC (República Argentina, Instituto Nacional de Estadística y Censos) (2002) Censo Nacional de Población y Viviendas 2001. Volumen I. Resultados provisionales. Total del País, Buenos Aires.

Iorio AF de, Barros MJ, Rendina A, Di Risio C, García A, Bargiela M, de Cabo L, Arreghini S, De Siervi M, Vella R (1997) Distribución de metales pesados en ácidos húmicos y fúlvicos de sedimentos del río Reconquista (Buenos Aires, República Argentina). Resúm Congr Internac sobre Aguas. Facultad de Derecho y Ciencias Sociales, Universidad de Buenos Aires, III. 38.

Joyce S (1997) Growing pains in South America. Environ Health Perspect 105:794–799.

Kreimer ED, Palacios DE, Ronco AE (1996) A proposal for dredging contaminated sediments at the Dock Sud Port, Argentina. Int Symp Coastal Ocean Space Utiliz COSU'96:435–444.

Kuczynski D (1984) Zooplancton (especialmente rotíferos) del río Reconquista Provincia de Buenos Aires). Physis 42B:1–7.

Kuczynski D (1991a) Atlas ecológico del arroyo Morón. Universidad de Morón, Buenos Aires.

Kuczynski D (1991b) Rotíferos del río Reconquista (Provincia de Buenos Aires, Argentina): Familia Brachionidae. An Soc Cient Argent 221:65–80.

Kuczynski D (1993) El Reconquista. Cronología de un río cercano. Letra Buena, Buenos Aires.

Kuczynski D (1994) Estudio ambiental de un curso de agua urbano altamente deteriorado por acción antropógena (arroyo Morón, provincia de Buenos Aires, Argentina). Rev Ecol Méd Salud Ambiental I:1–14.

Lacoste C, Collasius D (1995) Instrumentos de diagnóstico ambiental: índice de calidad de agua. Gerencia Ambiental 24:286–293.

Legèndre L, Legèndre P (1979) Ecologie numérique, vols I and II. Masson, Paris.

Loez CR (1995) Estudios limnológicos en el río Reconquista (Pcia. de Buenos Aires): relación entre parámetros biológicos y químicos, especialmente el impacto del Zn sobre la estructura del fitoplancton. Doctoral dissertation, Faculty of Exact and Natural Sciences, University of Buenos Aires.

Loez CR, Salibián A (1990) Premières données sur le phytoplancton et les charactéristiques physico-chimiques du rio Reconquista (Buenos Aires, Argentine). Une rivière urbaine polluée. Rev Hydrobiol Trop 23:283–296.

Loez CR, Topalián ML (1999) Use of algae for monitoring rivers in Argentina with special emphasis for the Reconquista river (region of Buenos Aires). In: Prygiel J, Whitton BA, Bukowska J (eds) Use of Algae for Monitoring Rivers, vol III. Agence de l'Eau Artois-Pircardie, pp 72–83.

Loez CR, Topalián ML, Salibián A (1995) Effects of zinc on the structure and growth of a natural freshwater phytoplankton assemblage reared in the laboratory. Environ Pollut 88:275–281.

Loez CR, Salibián A, Topalián ML (1998) Associations phytoplanctoniques indicatrices de la pollution par le zinc. Rev Sci Eau 3:315–332.

Loewy M, Kirs V, Carvajal G, Venturino A, Pechen de D'Ángelo AM (1999) Groundwater contamination by azinphos methyl in the Northern Patagonic Region (Argentina). Sci Total Environ 225:211–218.

Marbán L, López-Camelo LG de, Ratto S, Agostini A (1999) Contaminación con metales pesados en un suelo de la cuenca del río Reconquista. Ecología Aust 9:15–19.

Marteau SA, Alberino JC, Ripoli JL, Rosato ME (1998) Quality of water wells in an agricultural area in the city of La Plata, Argentina. Water Air Soil Pollut 106: 447–462.

Martínez JE, Salibián A (1995a) Spatial and temporal distribution of coliform group in the Reconquista river. Congr Int Assoc Theor Appl Limnology 1995:295 (abstr XXVI).

Martínez JE, Salibián A (1995b) Fecal pollution of surface water in an urban river. Proc 7th Int Symp Toxicity Assessment (ISTA 7) 1995:75.

Mateucci SD, Morello J, Rodríguez A, Buzai GD, Baxendale, C (1999) El crecimiento de la metrópoli y los cambios de biodiversidad: el caso de Buenos Aires. In: Mateucci SD, Solbrig OT, Morello J, Halffter G (eds) Biodiversidad y Uso de la Tierra. Conceptos y Ejemplos de Latinoamérica. EUDEBA-UNESCO, Buenos Aires, pp 549–580.

Momo F, Cuevas W, Giorgi A, Banchero M, Rivelli S, Taretto C, Gómez Vázquez A, Feijoo C (1999) Water quality of the Puelchense subaquifer in Luján (Argentina). In: Anagnostopoulos P, and Brebbia CA (eds) Water Pollution, vol V. Modelling, Measuring and Prediction. WIT Press, pp 493–501.

O'Farrell I, Lombardo R, Tezanos Pinto P de, Loez C (2002) The assessment of water quality in the lower Luján River (Buenos Aires, Argentina): phytoplankton and algal bioassays. Environ Pollut 120:207–218.

Olguín HF, Salibián A, Puig A (2000) Comparative sensitivity of *Scenedesmus acutus* and *Chlorella pyrenoidosa* as sentinel organisms for aquatic ecotoxicity assessment: studies on a highly polluted urban river. Environ Toxicol 15:14–22.

Olguín HF, Puig A, Loez CR, Salibián A, Topalián ML, Castañé PM, Rovedatti MG (2004) An integration of water physicochemistry, algal bioassays, phytoplankton, and zooplankton for ecotoxicological assessment in a highly polluted lowland river. Water Air Soil Pollut 155:355–381.

Olguín Salinas H, Puig A, Loez C, Salibián A, Topalián M, Castañé PM, Rovedatti M (1999) Evaluación ecotoxicológica integral de la calidad del agua del Río Reconquista: bioensayos, fisicoquímica, fito y zooplancton. Resúm 2a. Reunión SETAC-América Latina (Sección Argentina), TA-17.

Pereyra FX, Tchilinguirian P (2003) Problemas ambientales en el area metropolitana Bonaerense (AMBA): aspectos geológicos. In: Alsina G (Organizadora) Las aguas bajan turbias en la región metropolitana del Gran Buenos Aires. Ed Universidad Nacional de General Sarmiento—Ediciones Al Margen, Buenos Aires, pp 43–67.

Piazza A, Pérez-Lissarrague J, Barbado JL (2000) Guía práctica para el profesional fitoterapista. Ed Dunken, Buenos Aires.

Plá LE (1986) Análisis multivariado: método de componentes principales. Organización de los Estados Americanos. Programa Regional de Desarrollo Científico y Tecnológico. Monografía 27, Washington, DC.

Porta A (1996) Contaminación ambiental: uso de indicadores bioquímicos en evaluaciones de riesgo ecotoxicológico. Acta Bioquím Clín Latinoam 30:67–79.

Puig A (1997a) Crustáceos planctónicos y calidad del agua: caso del río Reconquista. Resúm II Congr Argent Limnología 1997:134.

Puig A (1997b) Gradiente espacial de calidad del agua y sus posibles bioindicadores: caso del Río Reconquista. Resúm Congr Internac sobre Aguas. Facultad de Derecho y Ciencias Sociales, Universidad de Buenos Aires III. 58.

Puig A, Salibián A (1999) Variaciones espaciotemporales del zooplancton en un río urbano contaminado (Río Reconquista, Buenos Aires). Resúm XIX Reunión Argent Ecología 1999:52.

Rand GM, Wells PG, McCarty, LS (1995) Introduction to aquatic toxicology. In: Rand GM (ed) Fundamentals of Aquatic Toxicology. Effects, Environmental fate, and Risk assessment. Taylor & Francis, Washington, DC, pp 3–67.

Ronco AE, Alzuet PR, Sobrero MC, Bulus Rossini G (1996) Ecotoxicological effects assessment of pollutants in the coastal region of the Gran La Plata, Province of Buenos Aires. Proc Int Conf Pollut Process Coastal Environ 1996:116–119.

Rovedatti MG, Loez CR, Topalián ML, Castañé PM, Salibián A, Olguin HF (2000) Monitoring of a polluted river (Reconquista, Argentina) based on physicochemical parameters and phytoplankton. Verh Int Verein Limnol 27:2743–2748.

Rovedatti MG, Castañé PM, Topalián ML, Salibián A (2001) Monitoring of organochlorine and organophosphorous pesticides in the water of the Reconquista River (Buenos Aires, Argentina). Water Res 35:3457–3461.

Salibián A (1992) Effect of deltamethrin on South American toad *Bufo arenarum* tadpoles. Bull Environ Contam Toxicol 48:616–621.

Salibián A (1999) Catálogo de las contribuciones de la Unlu referidas a la calidad del agua del Río Reconquista. Resúm Jorn Ciencia Tecnol Univ Nac Luján CB 73:51.

Saltiel GC (1997a) Situación ambiental de la cuenca hídrica del Río Reconquista. Tercera parte: el proyecto de saneamiento. Ing Sanit Amb 34:46–54.

Saltiel GC (1997b) Situación ambiental en la cuenca hídrica del Río Reconquista. Problemas y soluciones (primera parte). Ing Sanit Amb 31:45–55.

Saltiel GC, Romano L (1997) Situación ambiental en la cuenca hídrica del Río Reconquista (2ª parte). Ing Sanit Amb 32:29–36.

Topalián ML, Castañé PM (2003) Aplicación de índices químicos para la evaluación de la calidad del agua del río Reconquista. In: Alsina G (Organizadora) Las Aguas Bajan Turbias en la Región Metropolitana del Gran Buenos Aires. Ed Universidad Nacional de General Sarmiento—Ediciones Al Margen, Buenos Aires, pp 69–83.

Topalián ML, Loez CR (1995) Efecto del Zn sobre microalgas dulceacuícolas del Río Reconquista (Buenos Aires): bioensayo multiespecífico de invierno. Acta Toxicol Argent 2:6–9.

Topalián ML, Loez CR, Salibián A (1990) Metales pesados en el Río Reconquista (Buenos Aires): resultados preliminares. Acta Bioquím Clín Latinoam 24:171–176.

Topalián ML, Rovedatti MG, Castañé PM, Salibián A (1999a) Pollution in a lowland river system. A case study: the Reconquista River (Buenos Aires, Argentina). Water Air Soil Pollut 114:287–302.

Topalián ML, Castañé PM, Rovedatti MG, Salibián A (1999b) Principal component analysis of dissolved heavy metals in water of the Reconquista River (Buenos Aires, Argentina). Bull Environ Contamin Toxicol 63:484–490.

Topalián ML, Castañé PM, Salibián A, Romano LA, Rodríguez Capítulo A, Puig A (2001) Diversos enfoques sobre la situación del Río Reconquista. Agua Tecnología Tratamiento 136:38–42.

Tudino M (Coord) (2001) La contaminación del agua. In: Borthagaray JM, Fernández Prini R, Igarzábal de Nistal, San Román E (eds) Diagnóstico Ambiental del Area Metropolitana de Buenos Aires. FADU-UBA, Buenos Aires, pp 109–148.

UNDP, UNEP, WB, WRI (2000) World Resources 2000–2001. People and Ecosystems: The Fraying Web of Life. Elsevier, Washington DC.

Verrengia-Guerrero NR, Kesten EM (1998) Monitoreo ambiental y biológico en zonas costeras del Río de la Plata: relevancia de los contaminantes Cadmio y Plomo. Rev Mus Argentino Cs Nat, Volumen Extra (Ecotoxicología). Nueva Serie 147:1–12.

Vidal CM, Vitale AA, Viale AA (1993) Microorganismos degradadores de ácido naftalen-2-sulfónico. Rev Arg Microbiol 25:221–226.

Vigna MS, Alberghina J, del Mónaco SM, Galvagno MA (2002) *Prototheca zopfii* (Chlorophyta) capaz de utilizar "gas oil," registrada por primera vez en aguas contaminadas de Argentina. Darwiniana 40:45–50.

Villar C, Tudino M, Bonetto C, de Cabo L, Stripeiskis J, d'Huicque L, Troccoli O (1998) Heavy metal concentrations in the lower Paraná river and right margin of the Río de la Plata estuary. Verh Int Veren Limnol 26:963–966.

Wachs B (1998) A qualitative classification for the evaluation of the heavy metal contamination in river ecosystems. Verh Int Veren Limnol 26:1289–1294.

WHO (1993) Biomarkers and risk assessment: concepts and principles. IPCS Environmental Health Criteria Monographs 155. WHO, Geneva.

Zalazar RH (ed)(1996) Cuencas Hídricas. Contaminación. Evaluación de riesgo y saneamiento. Instituto Provincial del Medio Ambiente. Gobernación de la Provincia de Buenos Aires, Argentina.

Manuscript received October 9, 2003; accepted December 14, 2004.

Rev Environ Contam Toxicol 185:67–92

Paper Manufacture and Its Impact on the Aquatic Environment

J.P. Stanko, and R.A. Angus

Contents

I. Introduction

The history of paper can be traced back as early as 4000 B.C. when ancient Egyptians first utilized papyrus, from which the word paper was derived, as material on which to record information. Paper as we know it today first appeared in China around 105 A.D. and the first paper mill opened in Spain around 1009 A.D. The first North American paper mill was founded in 1690 in Philadelphia, Pennsylvania, where old cloth rags were recycled into paper products. By the late 1800s, paper was being produced from wood pulp, and by the early 1900s paper manufacturing had developed into a trade of significant economic importance. Pulp and paper manufacturing is now one of the largest industries in the world with annual sales of billions of dollars

Communicated by George W. Ware.

J.P. Stanko (✉)
United States Environ. Prot. Agency, Experimental Toxicology Div., Pharmacokinetics Branch, 109 T.W. Alexander Dr., MD-B143-01, Res Tri Park NC 27709.

R.A. Angus
Department of Biology, University of Alabama at Birmingham, 1300 University Boulevard, Birmingham, AL 35294, USA.

(USEPA 2002). The industry is the largest process water user in the United States, and only primary metal and chemical industries use more water throughout the rest of the world (USEPA 2002). An important consideration with respect to wastewater discharge is the process of pulping and the chemical input and output associated with pulp manufacturing. The average pulp mill uses between 16,000 and 46,000 L water/metric ton (t) of pulp processed (USEPA 2002), producing as much as 60 m^3 of wastewater/t of paper produced, and discharging nearly 16 million m^3 water/d (USEPA 2002). With such large amounts of wastewater being generated each day, it is not difficult to appreciate the potentially harmful impact that this industry presents to the environment. Although water is often recycled and chemical recovery systems reuse many process chemicals, the pulping process on the whole is chemical intensive and is the focus of much past and ongoing research and rulemaking.

The pulp and paper industry manufactures a vast array of products including newsprint, tissue, printing and writing paper, packaging and corrugated materials, and specialty papers. The manufacturing of pulp and paper is generally divided into two main processes: the production of pulp and the production of paper. Roughly half the pulp and paper making facilities are composed of mills involved exclusively in one of these two primary production processes. There are also facilities composed of integrated mills, which produce both pulp and paper, and converting facilities, which convert primary paper products to finished paper products such as envelopes and packaging material. According to a recent U.S. Census Bureau economic census, there are nearly 6000 paper-making establishments operating within the U.S., 514 of which are pulp and paper manufacturing facilities (USEPA 2002). Of these, 239 produce only paper, 39 produce only pulp, and the remainder are integrated facilities that produce both pulp and paper. The pulping process is the foremost environmental concern for the industry, and the production of pulp is the primary source of environmental pollutants and environmental impacts associated with pulp and paper manufacturing.

North America produces more than half the world's pulp. Pulp mills are geographically distributed throughout the southeast, northwest, northeast, and north central regions, where pulp trees are harvested from natural forests or tree farms. Mills in the northwest and southeast have been the focus of most effluent research in the U.S. because of the high concentration of mills and the subsequent high volume of effluent generated. However, pulp mills throughout Canada have also been the subject of much environmental scrutiny, and the bulk of data regarding the effects of exposure to pulp mill effluent stem from research conducted by Canadian investigators.

The effects of pulp and paper mill effluent (PME) have been a subject of extensive worldwide investigation over many decades (Larsson et al. 1988; Neuman and Karas 1988; Adams et al. 1992; Hodson et al. 1992; Galloway et al. 2003). Exposure to PME can elicit a variety of responses by affecting a wide range of biochemical and physiological processes. PME can induce morphological effects (Howell et al. 1980; Bortone et al. 1989; Cody and Bortone 1997), immunotoxic effects (Aaltonen et al. 1997; Fatima et al. 2001), reproductive dysfunction (Tana and Mikunen 1986; Leblanc et al. 1997; Van den Heuval et al. 2002), and disrupt liver function, leading to altered hepatic enzyme activity and sex steroid production (McMaster et al. 1992; Munkittrick et al. 1994; Gibbons et al. 1998; Alsop et al. 2003).

PME research has involved both terrestrial and aquatic species, including blue heron (Henshel et al. 1995), osprey (Elliot et al. 1998), eel (Pacheco and Santos 1999), and crayfish (King et al. 1999), but fish have been by far the most popular model species utilized in PME exposure studies. A variety of biomarkers have been utilized to assess the impact of PME on fish health, and changes have been observed at many different levels of biological organization. Table 1 lists various fish species utilized in effluent exposure studies and the effects of exposure.

Table 1. Various fish species utilized in pulp mill effluent exposure and effects.

Species	Dose (mg/L, mg/kg)	Exposure time	Effect
Catastomus commersoni	Effluent	Life	Reduced plasma sex steroid, decreased egg size and GSI, lowered response to GtH and GnRH, decreased steroid production, increased LSI and MFO activity
Coregonus clupeaformis	Effluent	Life	Decreased plasma cortisol
Cottus gobio	Effluent	Life	Changes in hepatic morphology
Danio rerio	0.03–0.09	7–28 d	Decreased hatchability and larvae survival
Esox lucius	Sediment	Life	Decreased stress response, degenerative pituitary corticotrophs
Gambusia affinis	Downstream of effluent	Life	Masculinization of females, precocious development of males
Heterandria formosa	Effluent	Life	Masculinization of females
Perca fluviatilis	3% effluent	Life	Reduced GSI
Perca flavescens	Sediment	Life	Decreased stress response, degenerative pituitary corticotrophs
Rutilius rutilius	3% effluent	Life	Reduced GSI
Salmo trutta	0.2%–0.5% effluent	90 d	Decreased egg numbers, reduced egg fertilization, fry mortality

GSI: gonadosomatic index; GtH: gonadotrophs; GnRH: gonadotropic-releasing hormone; LSI: liver somatic index; MFO: mixed-function oxidase.
Source: Kime (2001).

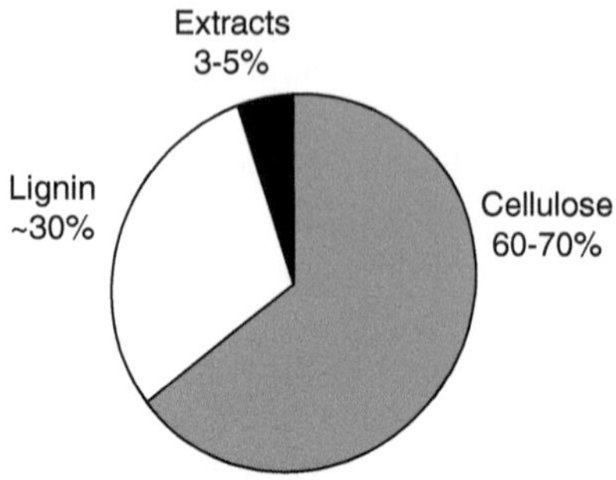

Fig. 1. Wood composition and relative amounts.

II. Paper Production Process

A. Wood Processing

Wood is comprised of cellulose, lignin, and extractives (Fig. 1). Cellulose consists of long-chain carbohydrates and is the primary component used in making paper. The remainder of the tree is made up of lignin, a phenolic polymer that acts as an adhesive to bind cellulose fibers, and extractives, which are plant hormones, resins, fatty acids, and other substances necessary for tree growth. The initial source of cellulose is known as *furnish*, and although it can consist of recycled or other post-consumer paper products, it is most commonly wood. Cellulose can also be obtained from nonwood sources, such as textiles, flax, hemp, and tobacco, but this type of furnish is generally used for specialty papers in only a small percentage of mills.

Trees are prepared for pulping by cutting, debarking, and chipping. The wood chips are then broken down to separate the cellulose from the noncellulose substances, and this raw material is transformed into pulp, a watery suspension of cellulose fibers. Depending on the pulping process utilized and the projected end product, the pulp can then be bleached to remove additional lignin and also to brighten the pulp for use in higher-quality paper products. Once bleached, the pulp suspension is applied to a screen where the water is drained off and the fibrous particles are left behind in a sheet to be finished into the final product. The overall sequence of processing events is wood preparation → pulping → pulp processing → bleaching → finishing. Although other "subprocesses" may be conducted, this is the general scheme. Most of the pollutants associated with papermaking are produced in the pulping and bleaching stages, during which the most chemical input and output occurs.

B. Pulping

The production of pulp is a major source of environmental impacts in paper making. Pulp production is achieved by one of three main processes: chemical, mechanical, or semichemical pulping. Each results in a different pulp end product varying in quality, which affects the overall quality of the finished product. Chemical pulping, the prevalent method, produces pulp that can be bleached to a greater extent due to its initially low lignin content, allowing for the production of the highest quality end product. Chemical pulps are produced by digesting (cooking) raw materials using kraft and sulfite processes. This method utilizes large amounts of chemicals to break down the wood in the presence of heat and pressure, yielding long fibers that are strong and stable. Furnish that is chemically pulped is frequently subjected to a bleaching stage because the pulp produced is brown in color.

Mechanical pulping, also called thermomechanical pulping (TMP), employs physical power to strip cellulose fibers off the wood with a grindstone. The power is commonly generated by hydroelectric or steam-driven systems that utilize recycled wastewater and burn wood scraps as a heat source. This method is used considerably less because the fibers are generally broken down and are thus of low strength and high in impurities. Wood that is mechanically pulped also retains most of the original lignin and usually undergoes further processing.

Semichemical pulping, or chemithermomechanical pulping (CMTP), uses a combination of chemical and mechanical processes. The wood is first softened by chemicals and then stripped mechanically, resulting in cellulose fibers that are generally stiff and of low quality. This process is used typically in the production of corrugated paper and paperboard products.

Kraft pulping and sulfite pulping are the two main types of chemical pulping in the U.S. The difference between the two methods is the chemicals used to digest the wood. Kraft processes produce a variety of pulps used mainly for packaging, high-strength papers and boards, and writing papers, because kraft pulps are much stronger than those produced by other processes. The characteristic strength of kraft fibers results from the use of digestive chemicals that are highly selective for wood constituents. Lignin removal is also high, allowing for bleaching without added pulp degradation. In the kraft process, wood is digested in an aqueous solution of sodium hydroxide and sodium sulfide in the presence of heat and pressure. This mixture, known as white liquor, dissolves the lignin and makes it soluble in the mixture. Sulfite pulping uses sulfurous acid and bisulfite ion to degrade the lignin of nonresinous softwoods as the predominant furnish. The end products can be bleached more easily than kraft pulps because they have less color, but are not as strong. Use of the sulfite pulping process has declined over time due to dependence on a particular type of wood and the absence of bark to achieve effective and efficient pulping. In 2000, less than 2% of the chemical market pulp capacity was sulfite produced (USDA 2001).

C. Pulp Processing and Bleaching

Pulp produced by chemical digestion, known as brownstock, contains separated wood fibers and black liquor, a solution of dissolved lignin solids and depleted pulping chemicals including organic extracts. Brownstock is subjected to washing to remove impurities and recycle the black liquor. Efficient washing is imperative to ensure thorough chemical recovery and to minimize organic waste from the large amounts of bleaching chemicals used. Excess dissolved organic compounds (lignins and extracts) left behind by inefficient washing can bind to bleaching chemicals, where they act as precursors to chlorinated organic compounds such as dioxins and furans. Common washing devices include rotary vacuum washers, diffusion washers, rotary pressure washers, and dilution/extraction washers. Once washed, the residual wastewater is passed through an evaporator and then into a recovery boiler while the "clean" brownstock is sent to the bleaching plant. The recovery boiler rids the wastewater of undigested wood chips, sand, and dirt, while concentrating the black liquor, and utilizes energy from combustion to convert the black liquor impurities into molten form, called smelt. The smelt is recausticized to convert sodium carbonate into active sodium hydroxide and sodium sulfide, and this "green liquor" is further chemically processed to regenerate the white liquor used in the digestion process.

The bleaching stage of paper manufacturing is the source of most organic environmental contaminants. During this stage, the pulp is subjected to alternative treatments with chlorine or chlorine derivatives and sodium hydroxide. The chlorine acts as an oxidizing agent to degrade the lignin and as a chlorinating agent to brighten the pulp, while sodium hydroxide treatment solubilizes the degraded lignin for extraction. During the chlorine and alkaline treatments, about 95% of the elemental chlorine is converted to chloride ions and the remainder is solubilized as chlorinated organic material. Effluent is then released into holding ponds where organic waste is stabilized by aeration and where microorganisms can metabolize additional excess organic waste before discharge into rivers or streams.

Elemental chlorine was utilized as the primary chlorine agent until the early 1990s, when the presence of chlorinated organic material in pulp mill effluent became a public concern. In 1990, the USEPA developed the National Pollutant Discharge Elimination System to ensure effluent discharges from pulp mills using chlorine bleaching met applicable requirements of the Clean Water Act. Since then, many mills have switched to chlorine dioxide bleaching as well as bleaching with oxygen derivatives such as hypochlorite, ozone, and hydrogen peroxide. The use of chlorine dioxide in bleaching, known as elemental chlorine free (ECF) bleaching, produces 20%–90% less chlorinated organic substances (Solomon et al. 1993).

III. Environmental Impacts

A. Chemical Outputs

Chemical pulping generates a significant amount of chemical output due to the amount of chemical input required to digest the furnish. If chemical pulp is not efficiently washed, it leaves behind excess organic compounds that can increase the probability of toxic compound formation and the amount of toxic compounds produced during bleaching. Bleaching stages that use elemental chlorine produce the most chlorinated organic compounds whereas those that utilize chlorine dioxide or oxygen derivative chemicals produce the least. In most processes, chlorinated and organic compounds, nutrients, and metals are discharged as wastewaters in pulp mill effluent. Table 2 lists the compounds produced during the respective stages and their intended end fates. The potential impact that wastewater discharge has on the

Table 2. Stages of paper production and general by-products.

Stage	Effluent characteristics	End fate
Pulping	Black liquor	Recycle
	Natural and synthetic chemicals originating from wood	Washing
Washing	Wash wastewater	Recycle
	Brownstock	Bleaching
Bleaching	Wood products (BOD, COD, TSS)	Treatment in holding ponds or discharge
	Bleaching effluent containing delignification products and chlorinated organic compounds (AOX, EOX, VOC)	

BOD: biological oxygen demand; COD: chemical oxygen demand; TSS: total suspended solids, AOX: adsorbable organic halides; EOX: extractable organic halides; VOC: volatile organic compounds.

environment can be assessed by measuring various water quality parameters. Biochemical oxygen demand (BOD), chemical oxygen demand (COD), total suspended solids (TSS), and adsorbable and extractable organic halides (AOX and EOX, respectively), which include chlorinated organic compounds, are parameters commonly measured to determine the cleanliness of a body of water. These methods are effective in providing a general picture of water quality but do not identify specific contaminants. Pulping and wastewater treatment also produce a number of volatile organic compounds (VOC), such as chloroform, aldehydes, alcohols, phenols, ketones, terpenes, nutrients such as phosphorus and nitrogen, and heavy metals, which are used as complexing agents to remove other transition metals present within the wood. Because VOCs are released predominantly through air emissions and metal agents are not released at levels considered to be of significant environmental concern, neither is discussed further.

B. Biological Oxygen Demand and Chemical Oxygen Demand

Biological (or biochemical) oxygen demand is defined as the rate at which molecular oxygen is consumed by microorganisms during the oxidation of organic matter (Cairns et al. 1995). It is typically referred to as BOD, and the standard test measures the dissolved oxygen consumed over a 5-d period (BOD_5). Values are reported as kg/t pulp product or as mg/L effluent. Chemical oxygen demand (COD) measures the oxygen consumed during oxidation by strong chemical oxidants when organic materials are not easily degraded by microorganisms. High BOD accelerates bacterial growth in the water, which consumes the oxygen and reduces available dissolved oxygen (dO_2) levels. Decreased dO_2 can impact living organisms by reducing metabolic activity or by increasing stress associated with an increase in energy expenditure to cope with physiological challenges caused by lack of oxygen and the presence of contaminants; this ultimately results in a reduction in biomass or total elimination of the biota. COD presents a potential hazard to the aquatic environment through the presence of oxidizing agents and heavy metals. BOD is a major pollutant output of furnish preparation, kraft pulping, and wastewater treatment whereas COD is a major output of kraft pulping, bleaching, papermaking, and wastewater treatment. The pollutants are introduced into rivers and streams primarily through pulp and treated effluents, and these effluents are generally high in BOD/COD, although levels have decreased significantly since the early 1980s. Reported values for BOD_5 range from 2 to 350 kg/t or 4 to 400 mg/L whereas COD levels can be as high as 200 kg/t or 1100 mg/L (Gaete et al. 2000; Thompson et al. 2001; QFIA 2001). Wastewaters with BOD levels of 100 mg/L or greater are considered exceedingly polluted with organic waste. Nitrates and phosphates present in effluent are bacterial nutrients that accelerate their growth and metabolism and contribute to elevated BOD and COD levels. Although BOD and COD do not indicate whether the substances present are potentially toxic, they imply a potential stress to the aquatic environment, as suggested by population changes resulting from sublethal exposure to bleached kraft mill effluent (BKME) (Stoner and Livingston 1978; McLeay and Brown 1979; Oikari et al. 1985).

C. Total Suspended Solids

Total suspended solids (TSS) are generally produced during furnish preparation and wastewater treatment. TSS can also be produced during the papermaking process when pulp is applied to screens and residual solids are released in the flowthrough. In a 1993 survey of 33 mills, the USEPA reported a TSS range from 0.24 to 9.79 kg/t in bleached kraft effluent (USEPA 2002), and the National Council for Air and Stream Improvement (NCASI) reported a mean of 5.5 kg/t in a survey of 41 mills in 1989 (Force 1995). While TSS are normally not directly toxic to organisms inhabiting the waterway, they do have an effect on water turbidity. High turbidity levels can impact feeding and growth of fish and other inhabitants, as well as respiration through impairment of gill function.

D. Adsorbable Organic Halides and Organic Pollutants

Pulp and paper mill effluent is composed of many different synthetic and naturally occurring substances. The presence of chlorinated compounds and other organic material produced during the pulping and bleaching stages of paper manufacture is of considerable importance. The amount of chlorinated organic material can be measured as adsorbable organic halides (AOX) or extractable organic halides (EOX). AOX is determined by measuring the approximate amount of organically bound chlorine that adsorbs to activated charcoal whereas EOX measures organic compounds extracted with a nonpolar solvent. Examples include, but are not limited to, phenols, guaicacols, polyaromatic hydrocarbons, dioxins, and furans. Amounts of AOX produced during these stages of paper production range from 1.2 kg/t to 3.6 kg/t (Kallqvist et al. 1989; Lancaster et al. 1996; Barraoca et al. 2001). Concentrations of AOX in effluents range from as low as 10 μg/L to as high as 100 mg/L (Hodson et al. 1992; Tyler et al. 1998; Gaete et al. 2000). Although studies have shown that no correlations exist between AOX/EOX concentrations and biological responses (O'Connor and Nelson 1993; Robinson et al. 1994), these values provide an estimate of the chlorinated organic constituents of pulp mill effluent.

Of particular environmental importance is the production of dioxins and furans generated when elemental chlorine used in the bleaching processes reacts with aromatic organic precursors that are present in the wood. Studies of humans exposed to high levels of dioxins and furans have documented temporary adverse effects on the liver, immune system, reproduction, and development, as well as cardiovascular disorders and carcinogenic effects (Arehart et al. 2004; Valic et al. 2004; Steenland et al. 2004).

Harbor seals in British Columbia, taken immediately after a reduction in pulp mill discharge, contained detectable levels of polychlorinated dibenzo-*p*-dioxins (PCDD) and furans (PCDF) and polychlorinated biphenyls (PCB) in their blubber (Addison et al. 2004). Harbor seals are good indicators of contaminant levels in coastal food chains because they are territorial predators at the top of the food chain. The highest levels were observed in seals inhabiting the Strait of Georgia, which receives effluent from six coastal pulp mills. This research group conducted no toxicity studies but other studies indicate that high levels of PCDD/F and PCB in seals may result in a reduction of immunocompetence (Ross et al. 1995). The authors state that decreased immunocompetence was not observed or was masked by other factors. Regardless, the detection of PCDD/F and PCB in these seals raises the concern for accumulation in other organisms exposed to pulp mill effluent.

Dioxin contamination associated with paper and pulp mill effluent has been reported in both North American and European countries (Abbott and Hinton 1996; Kostamo et al. 2000; Shelby and Mendonca 2001; Harris et al. 2003; Ross et al. 2004). The ability of dioxins to impart such a broad range of toxic effects prompted government agencies in Canada and the U.S. to establish guidelines in an effort to reduce and ultimately eliminate dioxin release in effluents.

IV. Aquatic Toxicology

A. Morphological Effects

The first reports of morphological alterations related to pulp and paper mill effluent exposure came in 1980 when Dr. Mike Howell observed a population of masculinized female mosquitofish (*Gambusia holbrooki*) inhabiting Eleven Mile Creek, a stream below the effluent discharge of a paper mill in Escambia County, Florida (Howell et al. 1980). Mosquitofish belong to the family Poeciliidae, in which males possess a modified anal fin called a gonopodium that is used to transfer sperm to the female during copulation. The authors observed that the entire female population inhabiting this stream exhibited this normally male characteristic. Shortly thereafter, masculinized females of two other poeciliids, the least killifish (*Heterandria formosa*) and the sailfin molly (*Poecilia latipinna*), were observed in the Fenholloway River in Taylor County, Florida (Drysdale and Bortone 1989).

Many years later, masculinized females of these same species were observed in Rice Creek, a tributary of the St. Johns River in Putnam County, Florida, which receives paper mill effluent (Bortone and Cody 1999). Drysdale and Bortone (1989) were able to reproduce this masculinization in the laboratory, although to a lesser extent, using effluent collected from Eleven Mile Creek. In addition to anal fin elongation, they reported several other morphologically masculine alterations including preanal length, dorsal and pelvic fin height, interorbital width, and eye diameter. Kovacs et al. (1995) showed that fathead minnows (*Pimephales promelas*) chronically exposed (275 d) to 5%, 10%, and 20% concentrations of effluent collected from a kraft mill in western Canada exhibited male skewed sex ratios on the basis of secondary sex characteristics, and Larsson et al. (1988) found significantly male-biased embryos in eelpout (*Zoarces viviparous*) inhabiting the Baltic Coast near a kraft pulp mill. Variations in the expression of sexual characteristics in fish are not uncommon (Bortone and Davis 1994), but altered sex ratios indicate the presence of chemicals in the effluent that can elicit disturbances during sexual development and affect the survival of inhabitant species.

Because PME is a complex mixture of synthetic and naturally occurring compounds, the exact cause of these alterations is unknown and identification of a single compound responsible for masculinization is highly improbable. The masculinizing capacity of PME suggests that the constituents responsible for these effects are androgenic or antiestrogenic in nature. It has long been known that exposure to androgenic hormones can induce masculinization in female poeciliids (Eversole 1941; Turner 1941 & 1942). Howell et al. (1980) suggested the hypothesis "that some chemical or combination of chemicals in the papermill effluent is exerting a strong androgenic effect on the Eleven Mile Creek population of *G. holbrooki*." This hypothesis was supported by Rosa-Molinar and Williams (1984), who demonstrated that the plant sterol β-sitosterol was implicated in the masculinization observed through aerobic and/or anaerobic transformation into an androgenic hormone.

This point of view was strengthened further when Denton et al. (1985) were able to masculinize female mosquitofish via exposure to β-sitosterol and stigmastanol in the presence of the common soil bacterium *Mycobacterium smegmatis*. Exposure to β-sitosterol has also been shown to alter the reproductive status of goldfish by decreasing levels of plasma reproductive steroids in both males and females (MacLatchy et al. 1997). These researchers identified a correlation between decreased steroid levels and a decrease of pregnenolone, suggesting the possibility that β-sitosterol could reduce the capacity of gonadal steroid biosynthesis by interfering with cholesterol availability or by disruption of side chain cleavage by P-450_{scc}. These observations suggest that although compounds produced as a direct result of pulping and bleaching may have an impact on the well-being of aquatic wildlife, the possibility of bacterial transformation of plant sterols must also be considered.

Although the production of androgenic substances through bacterial transformation is a reasonable hypothesis, it was not until recently that any specific androgens had been identified in paper mill effluent. Jenkins et al. (2001) detected the presence of androstenedione in Fenholloway River water containing PME. The investigators confirmed the likelihood of modification of phytosterols into androgenic compounds, but also suggest that androgens may be supplied directly as plant products in tall oils. Parks et al. (2001) also detected the presence of substances in Fenholloway River water associated with androgenic activity as measured by androgen receptor (AR)-dependent transcriptional activation. In a study by Durhan et al. (2002), androstenedione was detected in Fenholloway River water but not in the fractions that exhibited androgenic activity in cell-based assays. They proposed that androstenedione is not associated with the observed androgenic activity. Orlando et al. (1999) hypothesized that female mosquitofish inhabiting the Fenholloway River are masculinized by accumulation of testosterone resulting from inhibition of aromatase, the enzyme that converts testosterone into estradiol. However, they found that ovarian and brain aromatase activity was actually elevated in these fish, possibly in response to increased circulating androgens, which supports the hypothesis of masculinization by one or more androgenic constituent of the river water (Orlando et al. 2002).

The bacterial transformation hypothesis was strengthened by the identification of androstenedione and, more importantly, progesterone (2.4 and 155 nM, respectively) in the sediment of the Fenholloway River at concentrations considerably higher than in the water column (Jenkins et al. 2003). Sediment microbes could transform the progesterone, an androgen precursor, into androstenedione and other bioactive steroids. The source of progesterone is believed to be pine pulp-derived phytosterols present in the mill effluent. This idea is justified in the same study by the identification of low levels of progesterone (0.3 nM) in the sediment of a comparison stream, which is attributed to the natural breakdown of plant material. *Mycobacterium smegmatis* incubated in the presence of progesterone produced 17α-hydroxyprogesterone, androstenedione, and also androstadienedione (Jenkins et al. 2004). Reanalysis of Fenholloway sediment revealed androstadienedione (4.0 nM), which exhibited activity similar to androstenedione in AR-mediated transcriptional assays.

Figure 2 illustrates the proposed *in vitro* production of androgens from progesterone by *M. smegmatis*. The discovery of androstadienedione may be significant in that androstadienedione has been shown to have an inhibitory effect on aromatase and may irreversibly inactivate the enzyme (Covey and Hood 1982). Reducing or eliminating aromatase activity would prevent females from

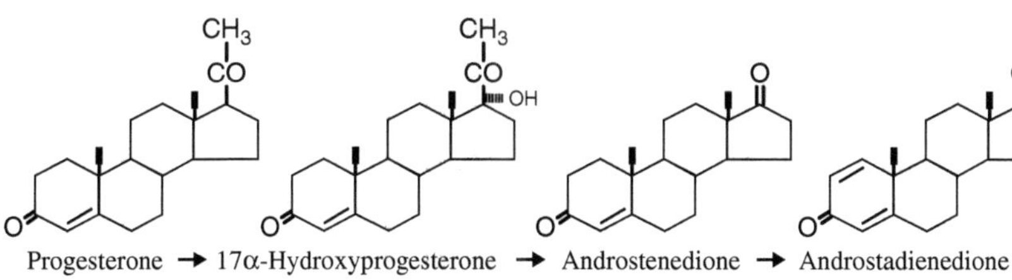

Fig. 2.

converting androgens into estrogens and could have a masculinizing, or defeminizing, effect. Androstenedione is more rapidly aromatized than androstadienedione and the results of the study by Orlando et al. (see above) suggest that the inhibitory effect of androstadiendione on aromatase in Fenholloway River female mosquitofish is minimal. This is supported by studies conducted in our laboratory where no change in VTG expression was observed in female mosquitofish exposed to androstadienedione concentrations as high as 500 nM by short-term static renewal method (Stanko 2005). If exposure to androstadienedione resulted in the inhibition of aromatase, an expected effect would have been a decrease in VTG expression. Thus, it is possible that high levels of androstenedione, by providing excess substrate, could account for the elevated aromatase activity.

Our laboratory has been unable to induce gonopodial development in female mosquitofish through short-term static exposure to androstenedione at concentrations measured in Fenholloway River water, but morphological masculinization has been observed within 2–3 weeks at concentrations 3 log doses higher (Stanko 2005). Although androstadienedione levels in the Fenholloway River water column have not been determined, our laboratory has also been able to induce morphological masculinization in female mosquitofish through short-term static renewal exposure to 50 and 500 nM androstadienedione (Stanko 2005). While these concentrations are likely much higher than in Fenholloway River water, these data suggest that androstadienedione almost certainly contributes to the morphological masculinization observed in Fenholloway River female mosquitofish. Steroids are very hydrophobic in nature, and therefore organisms inhabiting steroid-containing waters are susceptible to bioaccumulation of these compounds (Lai et al. 2002). Exposure to high levels of androstenedione, and also androstadienedione, may provide insight into the possible effects following long-term exposure.

B. Hepatic Dysfunction

The liver is the primary detoxification organ in all vertebrates and is therefore often used as an indicator of the sensitivity of an organism to environmental pollutants. Because of the lipophilic nature of many environmental organic pollutants, the potential for bioaccumulation in the liver and subsequent morphological damage and disruption of enzyme activity is evident. The liver is responsible for the production of the cytochrome P-450 superfamily of enzymes and their conjugates, as well as a number of other metabolically active enzymes. These enzymes play a major role in the synthesis and deactivation of many endogenous chemicals, including steroid hormones, as well as the detoxification of xenobiotics. The complexity of the mechanisms occurring within the liver presents an opportunity for the alteration of several biochemical pathways, and the wide variety of responses that result from such alterations has made the liver a focus of numerous toxicological studies. Therefore, the liver is frequently used as a bioindicator of the presence of pollutants in PME, primarily through effects on enzyme function, steroidogenesis, and vitellogenesis.

Mixed-function oxidases (MFO) are a versatile class of P-450-dependent hepatic enzymes involved in the oxidative conversion of endogenous lipophilic substrates into more hydrophilic compounds. Because of the normally endogenous nature of MFO substrates, these enzymes are usually present at relatively low activity. In the presence of environmental pollutants, activity of MFO enzymes is increased to rapidly deactivate these chemicals for excretion. MFO activity, typically measured through induction of ethoxyresorufin *O*-deethylase (EROD), is a sensitive biomarker of contaminant exposure, and induced MFO activity in fish is often measured as an indicator of exposure to PME (Lehtinen et al. 1990; Parrott et al. 2000; Kovacs et al. 2002; Galloway et al. 2003). Although increases in MFO activity are associated with effluent exposure, no direct link between mechanisms of effluent-induced activity and adverse effects has been demonstrated (Hodson et al. 1996). The inability to describe a direct link may likely be the result of the sensitivity of these enzymes, such that the induction detected precedes more serious aberrations.

Although a direct relationship between exposure and effects has not been established, significant induction of MFO indicates exposure to a toxicant and suggests the potential for adverse effects through other hepatic processes. The liver is also a primary site of steroidogenesis and steroid deactivation. Because many of the substrates of MFO enzymes have the same planar ring structure as hormones, it is not surprising to observe endocrine effects associated with elevated MFO activity. Dioxins and furans were detected in white sucker inhabiting the receiving areas of multiple Canadian pulp mills. In addition to elevated MFO activity and increased liver size, distinct alterations to reproductive fitness were also observed, including delayed sexual maturity and reduced GSI (Munkittrick et al. 1994). White sucker that inhabited the St. Maurice River, Canada, where the source of the only major industry discharge is a pulp mill, had elevated EROD activity and were found to be contaminated with chlorinated dioxins, furans, phenols, guaicacols, and veratrols (Gagnon et al. 1995). These fish also exhibited changes in various reproductive parameters, including decreased GSI in females (Table 3). The differences in age and size at maturity were attributed to a nutrient gradient resulting from an increased load of organic material and therefore a greater transfer of nutrients to the fish. Although growth appeared to be enhanced, the investigators reported marked differences in reproductive response. Soimasuo et al. (1998) reported an accumulation of chlorophenolics and resin acids in the bile of whitefish exposed *in situ* to treated bleached kraft mill effluent (BKME) concentrations greater than 3.5%. Whitefish utilized in the study also exhibited reductions in plasma testosterone and estradiol following exposure to 3.5% treated effluent.

Chemical contaminants identified in previous studies are not the only compounds present in pulp and paper mill effluent that have the ability to induce reproductive effects and modify endocrine processes. There are a number of both natural and anthropogenic constituents of PME associated with MFO activity that can alter reproductive homeostasis. Many chemicals that induce these hepatic enzymes may also be estrogenic or antiestrogenic, including a range of natural plant steroids with the potential for both estrogenic and androgenic activity. Changes

Table 3. Trends in biochemical and reproductive characteristics of white sucker in the Gastineau (reference site) and St. Maurice Rivers.

Characteristics	Gastineau River	St. Maurice River
MFO induction	No change	Increase
Age at maturity	Decrease	No change
Size at maturity	Decrease	Increase
Gonad size		
Male	Increase	Increase
Female	Increase	Decrease
Fecundity	Increase	No change

Source: Gagnon et al. (1995).

in circulating hormones may impart an effect on steroid production by the hypothalamic–pituitary system. Each of these responses leads to the consideration of the role of PME components in the reproduction and development of aquatic species.

C. Reproductive Effects

In addition to morphological effects, many reproductive effects have been described following exposure to PME. Its effects on reproductive impairment in fish include decreased gonadal size (Sandstrom et al. 1988; Gagnon et al. 1995), alterations of secondary sex characteristics (Howell et al. 1980; Bortone et al. 1989; Bortone and Cody 1999), and delayed sexual maturity (Munkittrick et al. 1992). Mummichog (*Fundulus heteroclitus*) living downstream from a bleached kraft pulp mill displayed delayed gonadal maturation and had smaller eggs and increased GSI (Leblanc et al. 1997). Observations near a Swedish pulp mill at the Bothnian Sea revealed lower relative gonad size and an increase in rate of growth in perch (*Perca fluviatalis*) (Sandstrom and Neuman 2003). Although there was an increase in rate of growth, the authors indicated this should not be interpreted as a positive response, as maturation of fish from the experimental site was delayed compared to those of the reference site. Our laboratory has observed a decrease in growth coupled with reductions in reproductive indicators such as gonadosomatic index (GSI) and circulating vitellogenin (VTG) in *G. affinis* following exposure to androstenedione as well as a significant negative association in overall growth and in the number of developing embryos per female following exposure to androstadienedione (Stanko 2005). This characteristic has been observed in female mosquitofish inhabiting the Fenholloway River. McNatt (McNatt 2002) indicated that Fenholloway females were significantly smaller than those inhabiting a reference site that did not receive effluent from a pulp mill and that pregnant Fenholloway females carried fewer developing embryos than their reference site counterparts, which could be attributed to their smaller size. Additionally, there was a positive association between fish size and number of developing embryos in females from the reference site. White suckers (*Catostomus commersoni*) exposed to BKME displayed increased ovarian follicular apoptosis and elevated heat shock protein 70 (HSP-70), indicating altered reproductive homeostasis and hormone signaling (Janz et al. 1997). In a later study, Janz et al. (2001) observed follicular apoptosis, decreased GSI, and elevated HSP-70 in white sucker exposed to BKME, and also determined that these parameters recovered following improvements made to the mill to reduce or eliminate process chemicals. Although a correlation between ovarian follicular apoptosis and elevated expression of HSP-70 has not been established, HSP-70 expression has been utilized as an indicator of environmental stress in aquatic species (Iwama et al. 1998; Feder and Hofmann 1999).

D. Endocrine Disruption

The issue of endocrine disruption and the existence of chemicals within the environment with the potential to disrupt endocrine function have rapidly been gaining attention. Endocrine disruptors (EDCs) are exogenous agents that mimic hormones in humans and animals with the potential to disrupt the normal function of an animal's endocrine system. EDCs can interfere with the critical timing events in growth and development through interactions with various steroid receptors, functioning as either agonists or antagonists (Kelce et al. 1995; Vonier et al. 1996; Crain and Guillette 1997). Regulation of reproduction in fish is generally controlled by the hypothalamic–pituitary–gonadal system. This system acts on external cues that signal a release of hormones from the hypothalamus (gonadotropic-, corticotropic-, and thyrotropic-releasing hormones; GnRH, CRH, and TRH, respectively) that then cause the release of hormones from the pituitary, primarily gonadotrophs (GtH). GtH largely regulate steroidogenic activity in the gonads, which can be affected by both gonadal and nongonadal factors. Steroidogenic activity in the gonads initiates secondary sexual characteristics and reproductive function and development. This activity is under both positive and negative feedback control based on stages of development and external signals and can be affected by exogenous contaminants. One particularly important process that can be hindered by endocrine disruptors is vitellogenesis. Vitellogenin (VTG) is a precursor yolk protein produced in the liver in response to circulating estrogen (Lazier and MacKay 1991). Oocyte growth is dependent upon circulating VTG, which is regulated by estradiol, which in turn is regulated by GtH (Wallace and Selman 1981). VTG expression has become a useful biomarker in assessing the impact of environmental contaminants on wildlife because of the number of mechanisms by which it can be modulated, particularly by circulating steroid hormone levels.

Studies investigating the properties of sex steroid-binding protein (SBP) in PME-exposed white sucker (*Catastomus commersoni*) and longnose sucker (*C. catostomus*) reported increased relative binding affinity for testosterone and estrogen (Pryce-Hobby et al. 2002). Testosterone-binding capacity was increased in prespawning female white sucker, and both estrogen- and testosterone-binding capacity were increased in late vitellogenic female longnose sucker. The authors state that SBP ligands are rapidly accumulated in the liver, particularly in exposed fish, but not in the bile, suggesting that these ligands are not excreted or are modified before biliary excretion so that they no longer have the ability to bind SBP. They also suggest that the accumulation of other compounds present in the effluent that do not bind SBP may be excreted in such high concentrations that they prevent the excretion of endogenous steroids.

Effects on binding capacity of SBP as a result of PME exposure may impart an effect on circulating steroid levels, resulting in the disturbance of natural steroid processes. Sepulveda et al. (2001) reported that largemouth bass (*Micropterus salmoides*) inhabiting effluent-dominated streams exhibited lower circulating levels of VTG, decreased plasma 11-ketotestosterone and 17β-estradiol concentrations, and elevated MFO activity. In various effluent concentration studies, these investigators determined that bass exposed to as low as 20% effluent exhibited declines in circulating sex steroids coupled with declines in GSI and ovarian development in females. These researchers suggest that the reproductive effects observed are a result of inhibition of overall reproductive function, rather than receptor-mediated activity, because of the nature of the effects observed. The reduction of circulating VTG in this study implies that PME has an antiestrogenic or androgenic effect, which is supported by the absence of VTG induction in male bass. This hypothesis is also supported by the observation in our laboratory of decreased circulating VTG in female mosquitofish exposed to androstenedione (Stanko 2005). However, Tremblay and Van der Kraak (1999) found elevated levels of VTG in immature rainbow trout exposed to BKME. They proposed the possibility of β-sitosterol binding to the trout estrogen receptor to induce VTG expression, as similar results have been shown in goldfish exposed to β-sitosterol. This finding would imply an estrogenic effect of β-sitosterol, which contrasts with Howell's observations (see earlier). The authors address this inconsistency and suggest these responses are caused by other compounds present in BKME.

Juvenile whitefish exhibited VTG gene expression following exposure to PMEs in Finland (Mellanen et al. 1999). VTG expression was demonstrated by measuring increased mRNA levels using Northern blot analysis, which further supports the argument that PME contaminants exhibit an estrogenic capacity. However, male rainbow trout exposed to a New Zealand PME showed no induction of VTG (Van den Heuval et al. 2002). This observation supports the hypothesis that PME is androgenic, but the authors suggest other possible reasons for lack of VTG induction, including species sensitivity and timing of exposure. It is likely that these varying responses are a result of multiple contaminants or impacts on different biochemical pathways. Alterations have been demonstrated as early in the reproductive process as the hypothalamic–pituitary axis. Van der Kraak et al. (1992) demonstrated that plasma GtH-II levels were considerably lower in male and female prespawning white sucker exposed to BKME than those from the reference site. Female fish at the BKME site failed to ovulate following injection of synthetic gonadotropin-releasing hormone (GnRH) whereas 100% of those at the reference site ovulated within 6 hr. Reduced testosterone levels in both sexes of fish at the BKME site were increased following treatment with GnRH analogue. These observations indicate that there are many sites in the hypothalamus–pituitary–gonad axis at which constituents of PME can induce an effect and suggest that additional effects can be produced at other locations in the reproductive development pathway. It is evident from these examples that the mechanisms by which VTG expression can be modulated are unclear and that more research is necessary to determine the nature of the effects observed. It is also clear that these processes are likely altered by multiple compounds with the ability to act on multiple pathways.

V. Discussion

The effluent produced as a result of chemical processing during pulp and paper manufacture contains both toxic and endocrine disrupting constituents. The effects of PME on aquatic organisms and identification of the constituents of PME responsible for these effects have been the topics of multiple investigations for many years. The accumulation of chlorinated and polychlorinated compounds, such as dioxins and furans, in the environment and in humans and wildlife is of significant environmental concern. These compounds are products of pulp and paper mills as well as a number of other industrial processes, and they have been the focus of a considerable amount of research examining their carcinogenicity, neurological and behavioral effects, and effect on reproductive dysfunction in both humans and animals (for general reviews, see Fletcher and McKay 1993; Kogevinas 2001; Mendola et al. 2002; Nilsson and Hakansson 2002).

The detection of chlorinated compounds in pulp and paper mill effluent prompted the U.S. Environmental Protection Agency (USEPA) to regulate dioxins in the paper industry and to develop effluent limitation guidelines and standards to reduce dioxin contamination in 1990. These guidelines led to the establishment of the USEPA Cluster Rule regulating both air and water emissions from pulp and paper mills (USEPA 1997) (The Cluster Rule can also be viewed at http://www.epa.gov/waterscience/pulppaper/cluster.html). Effluent discharge limits were placed on dioxins, furans, AOX, BOD_5, COD, TSS, and many chemicals commonly found in effluent. The USEPA states that the Cluster Rule will achieve 96% reduction in dioxins and furans and 99% reduction of chloroform in wastewaters. In conjunction with the Cluster Rule, the EPA released the Permit Guidance Document (USEPA 2000), that provides guidance to permit writers in developing National Pollutant Elimination Discharge System (NPEDS) permit requirements. In the document, the EPA provides effluent limitation guidelines describing the maximum discharge amounts for respective pollutants. Table 4 indicates effluent discharge limits for some of the pollutants addressed within this text. Canada also became aware of hazards presented by PME and developed the Pulp and Paper Effluent Regulations and Fisheries Act (Environment Canada 1992), that placed strict limitations on effluent discharge. Amendments were made in 2004 (Canada Gazette 2004) to update regulations and further ensure compliance.

In response to these guidelines, public awareness groups became active supporters of totally chlorine free (TCF) and elemental chlorine free (ECF) pulp bleaching processes, and many mills undertook measures to change their bleaching

Table 4. U.S. Environmental Protection Agency (EPA) Effluent Limitations Guidelines: New Source Performance Standards (NSPS).

Pollutant	1-d maximum discharge	Point of compliance
TCDD	Below minimum specified limits	Bleached plant effluent
TCDF	31.9 pg/L	Bleached plant effluent
AOX	0.476 kg/kkg	Final effluent
BOD	4.52 kg/kkg	Final effluent
TSS	8.47 kg/kkg	Final effluent
pH	5.0–9.0 at all times	Final effluent

TCDD: 2 3 7 8 Tetrachlorodibenzo-p-dioxin; TCDF: 2 3 7 8 Tetrachlorodibenzofuran. NSPS is defined as conventional, toxic, and nonconventional pollutants at a new source, direct discharger.
Source: USEPA 2000.

methods. TCF processes use oxygen and oxygen derivatives for delignification and bleaching and ECF processes utilize nonelemental chlorine derivatives, such as chlorine dioxide. These types of pulp and paper processing have been shown to reduce the number of chlorinated organic compounds in PME (Bankey et al. 1995; Martel et al. 1996). In a 2000 progress report, the American Forestry and Paper Association reported decreases since 1988 in BOD, TSS, and AOX associated with changes in pulping and bleaching practices (American Forestry and Paper Assoc. 2000). McMaster et al. (1996) were able to demonstrate partial recovery of hormone levels following experimental effluent exposure, but white sucker exposed *in situ* still retained detectable effects. Reproduction in fish also returned to normal when exposed to effluent from the same plant following mill changes (Kovacs and Megraw 1996).

However, many of the constituents of PME are environmentally persistent and continue to impart toxicity even after modifications are made. One example can be observed in European eels (*Anguilla anguilla* L.) exposed *in situ* to three polluted areas of the Vouga River, Cacia, Aveiro, Portugal, 2 yr following the closing and cleanup process of a BKME facility. The eels exhibited elevated plasma cortisol, glucose, and lactate levels at one or more locations within 8 hr of exposure, indicating that the waters remain contaminated (Teles et al. 2004). Environmental persistence suggests that it may be some time before we can accurately measure the effects of these cleanup efforts.

Other improvements in effluent processing include modified or extended cooking, oxygen delignification, and activated sludge treatment (AST), a secondary treatment used in conjunction with TCF or ECF bleaching. AST involves the use of bacterial degradation or transformation of effluent components into nontoxic and/or nonbioactive by-products. Several microorganisms with the ability to metabolize various industrial pollutants, including chlorinated organic compounds, have been identified; these include *Bacillus* (Andretta et al. 2004), *Rubrivivax* (Yu and Mohn 1999), and *Zoogloea* (Yu and Mohn 2002), among others. Although AST has been shown to reduce BOD in effluents by 86%–95% and COD by 40%–80% (Kostamo and Kukkonen 2003), it can also lead to excessive growth of filamentous microorganisms and problems associated with their growth, such as sludge bulking and foaming (Smith et al. 2003).

Disruptions to bacterial communities can also lead to inactivation or elimination of other compound-removing populations, particularly those involved in the removal of resin acids. Resin acids are diterpenoid carboxylic acids present in softwood trees commonly used in papermaking. They can accumulate in sediments, presenting potential acute and chronic toxicity, and some of their biotransformation compounds that are resistant to further degradation can bioaccumulate (Liss et al. 1997). The bacterial communities utilized in AST facilities must be closely monitored and maintained for maximum efficiency and to reduce the potential for toxicity resulting from population disturbances. An alternative to AST is anaerobic treatment. While not as widely used as AST, anaerobic treatment has many potential advantages including lower sludge production, lower chemical consumption, and smaller land requirements because of smaller reactors and energy production (Thompson et al. 2001).

Summary

The pulp and paper manufacturing industry generates large quantities of wastewater and has been described as a significant pollutant of the aquatic environment for many decades. The majority of pulp mills are located in the U.S. and Canada, and these countries are therefore subject to the greatest environmental impact. Some form of toxicity following exposure to mill effluent has been documented in numerous species, both aquatic and terrestrial, and the compounds present in effluents can affect multiple physiological processes, including hepatic mechanisms, reproduction and development, and endocrine function. Although the effects of PME on wildlife have been known for some time, the detection of chlorinated organic compounds generated substantial concern to humans because of the potential risk of exposure to these highly toxic compounds through environmental persistence and bioaccumulation. Government regulations were imposed on the pulping industry in an effort to prevent further contamination and reduce the toxic potential of pulp mill effluent to both humans and wildlife.

Included among these regulations was the elimination of bleaching processes that utilized organic chlorine in favor of elemental chlorine free (ECF) and totally chlorine free (TCF) bleaching processes. There is little in the peer-reviewed literature describing the advantages of TCF over ECF, and many Internet sources claim that the differences between the two are small, although ECF is better economically. The USEPA has listed ECF as one of the best available technologies (BAT) within the Cluster Rule. However, because the mechanisms of MFO activity are unclear and reproductive effects have been observed following effluent exposure, and because some of these same effects have been reported in areas minimally compromised by PME, it is difficult to assess the true improvements that these processes are attaining. Other methods of bleaching continue to be tested and each has its own beneficial aspects. However, negative elements still seem to arise, either through chemical persistence, the formation of previously undetected compounds, or the conversion from one toxicant to another. Certainly, the elimination of elemental chlorine in pulping and bleaching processes will improve the quality of effluents with respect to chlorinated organics, but further research is necessary to determine the actual benefits of all processes and to determine how the toxicity that remains can be eliminated.

References

Aaltonen TM, Valtonen ET, Jokinen EI (1997) Immunoreactivity of roach, *Rutilus rutilus*, following laboratory exposure to bleached pulp and paper mill effluents. Ecotoxicol Environ Saf 38:266–271.

Abbott JD, Hinton SW (1996) Trends in 2,3,7,8-TCDD concentrations in fish tissues downstream of pulp mills bleaching with chlorine: short communication. Environ Toxicol Chem 15(7):1163–1165.

Adams SM, Crumby WD, Greeley MS Jr, Shugart LR, Saylor CF (1992) Responses of fish populations and communities to pulp mill effluents: a holistic assessment. Ecotoxicol Environ Saf 24:347–360.

Addison RF, Ikonomou MG, Smith TG (2004) PCDD/F and PCB in harbor seals (*Phoca vitulina*) from British Columbia: response to exposure to pulp mill effluents. Mar Environ Res 59(2):165–176.

Alsop D, Hewitt M, Kohli M, Brown S, Van der Kraak G (2003) Constituents within pulp mill effluent deplete retinoid stores in white sucker and bind to rainbow trout retinoic acid receptors and retinoid X receptors. Environ Toxicol Chem 22(12):2969–2976.

American Forestry and Paper Association (2002) Environmental, Health and Safety Principles. American Forestry and Paper Association,

Andretta CWS, Rosa RM, Tondo EC, Gaylarde CC, Henriques JAP (2004) Identification and characterization of a *Bacillus subtilus* IS13 strain involved in the biodegradation of 4,5,6-trichloroguaiacol. Chemosphere 55:631–639.

Arehart E, Giasson G, Walsh MT, Patterson H (2004) Dioxin alters the human low-density and very low-density lipoprotein structure with evidence for specific quenching of Trp-48 in apolipoprotein C-II. Biochemistry 43(26):8503–8509.

Bankey LA, Van Veld PA, Borton DL, LaFleur L, Stegeman JJ (1995) Responses of cytochrome P4501A in freshwater fish exposed to bleached kraft mill effluent in experimental stream channels. Can J Fish Aquat Sci 52:439–447.

Barraoca MJM, Seco IM, Fernandes PM, Ferreira LM, Castro JA (2001) Reduction of AOX in the bleach plant of a pulp mill. Environ Sci Technol 35(21):4390–4393.

Bortone SA, Cody RP (1999) Morphological masculinization in poeciliid females from a paper mill effluent receiving tributary of the St. Johns River, Florida, USA. Bull Environ Contam Toxicol 63:150–156.

Bortone SA, Drysdale DT (1981) Additional evidence for environmentally induced intersexuality in poeciliid fishes. Assoc Southeast Biol Bull 58:67.

Bortone SA, Davis WP (1994) Fish intersexuality as indicator of environmental stress. Bioscience 44(3):165–172.

Bortone SA, Davis WP, Bundrick CM (1989) Morphological and behavioral characters in mosquitofish as potential bioindication of exposure to kraft mill effluent. Bull Environ Contam Toxicol 43:370–377.

Canada Gazette (2004) Regulations amending the pulp and paper effluent regulations. SOR/2004-109. Canada Gazette Part II 138(10):118–178.
Cairns J Jr, Niederlehner BR, Smith EP (1995) Ecosystem effects: functional end points. In: Rand GM (ed) Fundamentals of Aquatic Toxicology: Effects, Environmental Fate, and Risk Assessment, 2nd Ed. Taylor & Francis, Washington, DC, p 591.
Cody RP, Bortone SA (1997) Masculinization of mosquitofish as an indicator of exposure to kraft mill effluent. Bull Environ Contam Toxicol 58:429–436.
Covey DF, Hood WF (1982) A new hypothesis based on suicide substrate inhibitor studies for the mechanism of action of aromatase. Cancer Res (suppl 8):3327s–3333s.
Crain DA, Guillette LJ Jr (1997) Endocrine-disrupting contaminants and reproduction in vertebrate wildlife. Rev Toxicol 1:47–70.
Denton TE, Howell WM, Allison JJ, McCollum J, Marks B (1985) Masculinization of female mosquitofish by exposure to plant sterols and *Mycobacterium smegmatis*. Bull Environ Contam Toxicol 35:627–632.
Drysdale DT, Bortone SA (1989) Laboratory induction of intersexuality in the mosquitofish, *Gambusia affinis*, using paper mill effluent. Bull Environ Contam Toxicol 43:611–617.
Durhan EJ, Lambright V, Butterworth BC, Kuehl OW, Orlando EF, Guillette LJ Jr, Gray LE, Ankley GT (2002) Evaluation of androstenedione as an androgenic component of river water downstream of pulp and paper mill effluent. Environ Toxicol Chem 21(9):1973–1976.
Elliott JE, Machmer MM, Henry CJ, Wilson LK, Norstrom RJ (1998) Contaminants in osprey from the Pacific Northwest: I. Trends and patterns in polychlorinated dibenzo-*p*-dioxins and -dibenzofurans in eggs and plasma. Arch Environ Toxicol Chem 35(4):620–631.
Environment Canada (1992) Fisheries Act: Pulp and Paper Effluent Regulations. SOR/92-269. Environment Canada, Ottawa.
Eversole WJ (1941) The effects of pregnenolone and related steroids on sexual development in fish, *Lebistes reticulatus*. Endocrinology 28:603–610.
Fatima M, Ahmad I, Siddiqui R, Raisuddin R (2001) Paper and pulp mill effluent-induced immunotoxicity in freshwater fish *Channa punctatus* (Bloch). Arch Environ Toxicol Chem 40:271–276.
Feder ME, Hofmann GE (1999) Heat-shock proteins, molecular chaperones, and the stress response: evolutionary and ecological physiology. Annu Rev Phys 61:243–282.
Fletcher CL, McKay WA (1993) Polychlorinated dibenzo-*p*-dioxins (PCDDs) and dibenzofurans (PCDFs) in the aquatic environment: a literature review. Chemosphere 26:1041–1069.
Force PT (1995) Environmental comparison of bleached kraft pulp manufacturing technologies. Duke University Environmental Defense Fund,
Gaete H, Larrain A, Bay-Schmith E, Baeza J, Rodriguez J (2000) Ecotoxicological assessment of two pulp mill effluents, Biobio River Basin, Chile. Environ Contam Toxicol 65:183–189.
Gagnon MM, Bussieres D, Dodson JJ, Hodson PV (1995) White sucker (*Catostomus commersoni*) growth and sexual maturation in pulp mill-contaminated and reference rivers. Environ Toxicol Chem 14(2):317–327.
Galloway BJ, Munkittrick KR, Currie S, Gray MA, Curry RA, Wood CS (2003) Examination of the responses of slimy sculpin (*Cottus cognatus*) and white Sucker (*Catastomus commersomi*) collected on the St. John River (Canada) downstream of pulp mill, paper mill, and sewage discharges. Environ Toxicol Chem 22(12):2898–2907.
Gibbons WN, Munkittrick KR, McMaster ME, Taylor WD (1998) Monitoring aquatic environments receiving industrial effluents using small fish species 2. Comparison between

responses of trout-perch (*Percopsis omiscomaycus*) and white sucker (*Catostomus commersoni*) Downstream of a Pulp Mill. Environ Toxicol Chem 17(11):2238–2245.
Harris ML, Wilson LK, Norstrom RJ, Elliott JE (2003) Egg concentrations of polychlorinated dibenzo-*p*-dioxins and dibenzofurans in double-crested (*Phalacrocorax auritus*) and pelagic (*P. pelagicus*) cormorants from the Strait of Georgia, Canada, 1973–1998. Environ Sci Technol 37(5):822–831.
Henshel DS, Martin JW, Norstrom R, Whitehead P, Steeves JD, Cheng KM (1995) Morphopmetric abnormalities in brains of great blue heron hatchlings exposed in the wild to PCDDs. Environ Health Perspect 103 (suppl 4):61–66.
Hodson PV, McWhirter M, Ralph K, Gray B, Thiveierge D, Carey J, Van Der Kraak G, Whittle DM, Levesque M (1992) Effects of bleached kraft mill effluent on fish in the St. Maurice River, Quebec. Environ Toxicol Chem 11:1635–1651.
Hodson PV, Efler S, Wilson JY, El-Shaarawi A, Maj M, Williams TG (1996) Measuring the potency of pulp mill effluents for induction of hepatic mixed-function oxygenase activity in fish. J Toxicol Environ Health 49:83–110.
Howell WM, Black DA, Bortone SA (1980) Abnormal expression of secondary sex characters in a population of mosquitofish, *Gambusia holbrooki*: evidence for environmentally induced masculinization. Copeia 4:676–681.
Iwama GK, Thomas PT, Forsyth RHB, Vijayan MM (1998) Heat shock protein expression in fish. Rev Fish Biol Fishes 8:35–56.
Janz DM, McMaster ME, Munkittrick KR, Van Der Kraak G (1997) Elevated ovarian follicular apoptosis and heat shock protein-70 expression in white sucker exposed to bleached kraft pulp mill effluent. Toxicol Appl Pharmacol 147:391–398.
Janz DM, McMaster ME, Weber LP, Munkittrick KR, Van Der Kraak G (2001) Recovery of ovary size, follicle cell apoptosis, and HSP-70 expression in fish exposed to bleached pulp mill effluent. Can J Fish Aquat Sci 58:620–625.
Jenkins R, Angus RA, McNatt H, Howell WM, Kemppainen JA, Kirk M, Wilson EM (2001) Identification of androstenedione in a river containing paper mill effluent. Environ Toxicol Chem 20(6):1325–1331.
Jenkins R, Wilson EM, Angus RA, Howell WM, Kirk M (2003) Androstenedione and progesterone in the sediment of a river receiving paper mill effluent. Toxicol Sci 73:53–59.
Jenkins R, Wilson EM, Angus RA, Howell WM, Kirk M, Moore R, Nance M, Brown A (2004) Production of androgens by microbial transformation of progesterone *in vitro*: a model for androgen production in rivers receiving paper mill effluent. Environ Health Perspect 112(15):1508–1511.
Kallqvist T, Carlberg GE, Kringstad A (1989) Ecotoxicological characterization of industrial wastewater-sulfite pulp mill with bleaching. Ecotoxicol Environ Saf 18(3): 321–336.
Kelce WR, Stone CR, Laws SC, Gray LE, Kemppainen JA, Wilson EM (1995) Persistent DDT metabolite p,p′-DDE is a potent androgen receptor antagonist. Nature (Lond) 375:581–585.
Kime DE (2001) Endocrine Disruption in Fish. Kluwer Academic Publishers. Boston MA. pp 320–323.
King HM, Baldwin DS, Rees GN, McDonald S (1999) Apparent bioaccumulation of Mn derived from paper-mill effluent by the freshwater crayfish *Cherax destructor*: the role of Mn oxidizing bacteria. Sci Total Environ 226:261–267.
Kogevinas M (2001) Human health effects of dioxins: cancer, reproductive and endocrine system effects. Hum Reprod Update 7(3):331–339.

Kostamo A, Medvedev M, Pellinen J, Hyvärinen H, Kukkonen JVK (2000) Analysis of organochlorine compounds and extractable organic halogen in three subspecies of ringed seal from northeast Europe. Environ Toxicol Chem 19(4):848–854.

Kostamo A, Kukkonen JVK (2003) Removal of resin acids and sterols from pulp mill effluents by activated sludge treatment. Water Res 37:2813–2820.

Kovacs TG, Megraw SR (1996) Laboratory responses of whole organisms exposed to pulp and paper mill effluents: 1991–1994. In: Servos MR, Munkittrick KR, Carey JH, Van der Kraak GJ (eds) Environmental Fate and Effects of Pulp and Paper Mill Effluents. St. Lucie Press, Boca Raton, p 459–472.

Kovacs TG, Gibbons JS, Tremblay LA, O'Connor BI, Martel PH, Voss RH (1995) The effects of a secondary-treated bleached kraft mill effluent on aquatic organisms as assessed by short-term and long-term laboratory tests. Ecotoxicol Environ Saf 31:7–22.

Kovacs TG, Martel Voss RH (2002) Assessing the biological status of fish in a river receiving pulp and paper mill effluents. Environ Pollut 118(1):123–140.

Lai KM, Scrimshaw MD, Lester JN (2002) Prediction of the Bioaccumulation Factors and Body Burden of Natural and Synthetic Estrogens in Aquatic Organisms in the River Systems. Sci Total Environ 289:159–168.

Lancaster L, Renard J, Yin C, Phillips RB (1996) The effects of alternative pulping and bleaching processes on product performance: economic and environmental concerns. U.S. EPA, Washington, DC.

Larsson A, Andersson T, Forlin L, Hardig J (1988) Physiological disturbances in fish exposed to bleached kraft mill effluents. Water Sci Technol 20:67–76.

Lazier C, MacKay ME (1991) Vitellogenin expression in teleost fish. In: Hochachka PW, Mommsen T (eds) Biochemistry and Molecular Biology of Fishes. Molecular Biology Frontiers, 2nd Ed. Elsevier, Amsterdam, pp 391–405.

Leblanc J, Couillard CM, Brethes JF (1997) Modifications of the reproductive period in mummichog (*Fundulus heteroclitus*) living downstream from a bleached kraft pulp mill in the Miramichi Estuary, New Brunswick, Canada. Can J Fish Aquat Sci 54: 2564–2573.

Lehtinen KJ, Kierkegaard A, Jakobsson E, Wandell A (1990) Physiological effects in fish exposed to effluents from mills with six different bleaching processes. Ecotoxicol Environ Saf 19(1):33–46.

Liss SN, Bicho PA, Saddler JN (1997) Microbiology and biodegradation of resin acids in pulp mill effluents: a minireview. Can J Microbiol 43(7):599–611.

MacLatchy D, Peters L, Nickle J, Van der Kraak G (1997) Exposure to β-sitosterol alters the endocrine status of goldfish differently than 17 β-estradiol. Environ Toxicol Chem 16(9):1895–1904.

Martel PH, Kovacs TG, Voss RH (1996) Effluent from Canadian pulp and paper mills: a recent investigation of their potential to induce mixed function oxygenase activity in fish. In: Servos MR, Munkittrick KR, Carey JH, Van der Kraak GJ (eds) Environmental Fate and Effects of Pulp and Paper Mill Effluents. St. Lucie Press, Boca Raton, FL pp 401–412.

McLeay DJ, Brown DA (1979) Stress and chronic effects of untreated and treated bleached kraft pulp mill effluent on the biochemistry and stamina of juvenile coho salmon (*Onchorynchus kisutch*). J Fish Res Board Can 36:1049–1059.

McMaster ME, Portt CB, Munkittrick Dixon DG (1992) Milt characteristics, reproductive performance and larval survival and development of white sucker exposed to kraft mill effluent. Ecotoxicol Environ Saf 23:103–117.

McMaster ME, Munkittrick KR, Van der Kraak GJ, Flett PA, Servos MR (1996) Detection of steroid hormone disruptions associated with pulp mill effluent using artificial exposures of goldfish. In: Servos MR, Munkittrick KR, Carey JH, Van der Kraak GJ (eds) Environmental Fate and Effects of Pulp and Paper Mill Effluents. St. Lucie Press, Boca Raton, FL pp 425–438.

McNatt H (2002) Reproductive Fitness of Masculinized Female Eastern Mosquitofish (*Gambusia holbrooki*) from a River Receiving Pulp Mill Effluent. Masters thesis. University of Alabama at Birmingham, Birmingham, AL.

Mellanen P, Soimasuo M, Holmbom B, Oikari A, Santti R (1999) Expression of vitellogenin gene in the liver of juvenile whitefish (*Coregonus lavaretus* L.) Exposed to Effluents from Pulp and Paper Mills. Ecotoxicol Environ Saf 43(2):133–137.

Mendola P, Selevan SG, Gutter S, Rice D (2002) Environmental factors associated with a spectrum of neurodevelopmental deficits. Mental Retard Dev Dis Res Rev 8(3):188–197.

Munkittrick KR, McMaster MR, Portt CB, van der Kraak GJ, Smith IR, Dixon DG (1992) Changes in maturity, plasma sex steroid levels, hepatic mixed-function oxygenase activity, and the presence of external lesions in lake whitefish (*Coregonus clupeaformis*) exposed to bleached kraft mill effluent. Can J Fish Aquat Sci 49(8):1560–1569.

Munkittrick KR, Van der Kraak GJ, McMaster ME, Portt CB, van der Heuvel MR, Servos MR (1994) Survey of receiving-water-environmental impacts associated with discharges from pulp mills. 2. Gonad size, liver size, hepatic EROD activity and plasma sex steroid levels in white sucker. Environ Toxicol Chem 13(7):1089–1101.

Neuman E, Karas P (1988) Effects of pulp mill effluent on a Baltic coastal fish community. Water Sci Technol 20:95–106.

Nilsson CB, Hakansson H (2002) The retinoid signaling system: a target in dioxin toxicity. Crit Rev Toxicol 32(3):211–232.

O'Connor BI, Nelson S (1993) A study of the relationship between laboratory bioassay response and AOX content for pulp mill effluents. J Pulp Paper Sci 19:J33–J39.

Oikari A, Nikinimma M, Lindgren S, Lonn B (1985) Sublethal effects of simulated pulp mill effluents on the respiration and energy metabolism of rainbow trout (*Salmo gairdneri*). Ecotoxicol Environ Saf 9:378–384.

Orlando EF, Davis WP, Catabiano L, Bass D, Guillette LJ Jr (1999) Aromatase activity in the ovary of mosquitofish, *Gambusia holbrooki*, collected from the Fenholloway and Econfina Rivers, Florida. In: Ehormone Conference, Tulane University, New Orleans, LA.

Orlando EF, Davis WP, Guillette LJ Jr (2002) Aromatase activity in the ovary and brain of eastern mosquitofish (*Gambusia holbrooki*) exposed to paper mill effluent. Environ Health Perspect 110(suppl 3):429–433.

Pacheco M, Santos MA (1999) Biochemical and genotoxic responses of adult eel (*Anguilla anguilla* L.) to resin acids and pulp mill effluent: laboratory and field experiments. Ecotoxicol Environ Saf 42:81–93.

Parks LG, Lambright CS, Orlando EF, Guillette LJ Jr, Ankley GT, Gray LE Jr (2001) Masculinization of female mosquitofish in kraft mill effluent-contaminated Fenholloway River water is associated with androgen receptor agonist activity. Toxicol Sci 62(2):257–267.

Parrott JL, van der Heuvel MR, Hewitt LM, Baker MA, Munkittrick KR (2000) Isolation of MFO inducers from tissues of white suckers caged in bleached kraft mill effluent. Chemosphere 41(7):1083–1089.

Pryce-Hobby AC, McMaster ME, Hewitt LM, Van Der Kraak G (2002) The effects of pulp mill effluent on the sex steroid binding protein in white sucker (*Castostomus commersoni*) and longnose sucker (*C. castostomus*). Comp Biochem Phys C 134:241–250.

QFIA (Quebec Forest Industries Association) (2001) Production of the Environment Report. QFIA, Quebec.

Robinson RD, Carey JH, Solomon KR, Smith IR, Servos MR, Munkittrick KR (1994) Survey of receiving water environmental impacts associated with discharges from pulp mills 1. Mill characteristics, receiving water chemical profiles, and laboratory chronic toxicity tests. Environ Toxicol Chem 13:1075–1088.

Rosa-Molinar E, Williams CS (1984) Notes on the fecundity of an arrenhoid population of mosquitofish, *Gambusia holbrooki*. Northeast Gulf Sci 7(1):121–125.

Ross PS, DeSwart RL, Reijnders PJH, Van Loveren H, Vos JG, Osterhaus ADME (1995) Contaminant-related suppression of delayed-type hypersensitivity and antibody responses in harbor seals fed herring from the Baltic Sea. Environ Health Perspect 103: 162–167.

Ross PS, Jeffries SJ, Yunker MB, Addison RF, Ikonomou MG, Calambokidis JC (2004) Harbor seals (*Phoca vitulina*) in British Columbia, Canada, and Washington State, USA, reveal a combination of local and global polychlorinated biphenyl, dioxin, and furan signals. Environ Toxicol Chem 23(1):157–165.

Sandstrom O, Neuman E (2003) Long-term development in a Baltic fish community exposed to bleached pulp mill effluent. Aquat Toxicol 37:267–276.

Sandstrom O, Neuman E, Karas P (1988) Effects of bleached pulp mill effluent on growth and gonad function in Baltic coastal fish. Water Sci Technol 20:107–118.

Sepulveda MS, Reussler ND, Denslow ND, Holm SE, Schoeb TR, Gross TS (2001) Assessment of reproductive effects in largemouth bass (*Micropterus salmoides*) exposed to bleached/unbleached kraft mill effluents. Arch Environ Toxicol Chem 41:475–482.

Shelby JA, Mendonca MT (2001) Comparison of reproductive parameters in male yellow-blotched map turtles (*Graptemys flavimaculata*) from a historically contaminated site and a reference site. Comp Biochem Phys C Toxicol Pharmacol 129(3):233–242.

Smith NR, Yu Z, Mohn W (2003) Stability of a bacterial community in a pulp mill effluent treatment system during normal operation and a system shutdown. Water Res 37:4873–4884.

Soimasuo MR, Lappivaara J, Oikari OJ (1998) Confirmation of in situ Exposure of fish to secondary treated bleached kraft mill effluent using a laboratory simulation. Environ Toxicol Chem 17(7):1371–1379.

Solomon K, Bergmen H, Hugget R, Mackay D, McKague B (1993) A review and assessment of the ecological risks associated with the use of chlorine dioxide for the bleaching of pulp. Alliance for Environmental Technology, Washington, DC.

Stanko JP (2005) Reproductive and Developmental Effects of Bioactive Constituents of Pulp Mill Effluent on Female Mosquitofish, *Gambusia affinis*. Doctoral dissertation. University of Alabama at Birmingham, Birmingham, AL.

Steenland K, Bertazzi P, Baccarelli A, Kogevinas M (2004) Dioxin revisited: developments since the 1997 IARC classification of dioxin as a human carcinogen. Environ Health Perspect 112(13):1265–1268.

Stoner AW, Livingston RJ (1978) Respiration, growth, and food conversion efficiency of pinfish (*Lagodon rhomboides*) exposed to sublethal concentrations of bleached kraft mill effluent. Environ Pollut 17:207–217.

Tana J, Mikunen E (1986) Impact of pulp mill effluent on egg hatchability of pike (*Esox lucius* L.). Bull Environ Contam Toxicol 36:738–743.

Teles M, Santos MA, Pacheco M (2004) Responses of European eel (*Anguilla anguilla* L.) in two polluted environments: in situ experiments. Ecotoxicol Environ Saf 58:373–378.

Thompson G, Swain J, Kay M, Forster CF (2001) The treatment of pulp and paper mill effluent: a review. Biosour Technol 77:275–286.

Tremblay L, Van der Kraak G (1999) Comparison between the effects of the phytosterol β-sitosterol and pulp and paper mill effluents on sexually immature rainbow trout. Environ Toxicol Chem 18(2):329–336.

Turner CL (1941) Gonopodial characteristics produced in the anal fins of female *Gambusia affinis affinis* by treatment with ethyl testosterone. Biol Bull 80:371–383.

Turner CL (1942) A quantitative study of the effects of different concentrations of ethyl testosterone and methyl testosterone in the production of gonopodia in females of *Gambusia affinis*. Physiol Zool 15:263–280.

Tyler CR, Jobling S, Sumpter JP (1998) Endocrine disruption in wildlife: a critical review of the evidence. Crit Rev Toxicol 28(4):319–361.

USDA (2001) United States Paper, Paperboard, and Market Pulp Capacity Trends by Process and Location, 1970–2000. FPL-RP-602. USDA, Washington, DC.

USEPA (1997) EPA's Final Pulp, Paper, and Paperboard "Cluster Rule" Overview. EPA-821-F-97-016. Dept. of water, U.S. Environmental Protection Agency, Washington, DC.

USEPA (2000) Permit Guidance Document, Pulp, Paper and Paperboard Manufacturing, Point Source Category. 40 CFR §430. EPA-821-B-00-003. USEPA, Washington, DC.

USEPA (2002) Profile of the Pulp and Paper Industry, 2nd Ed. EPA/310-R-95-015.

Valic E, Jahn O, Papke O, Winker R, Wolf C, Rudiger WH (2004) Transient increase in micronucleus frequency and DNA effects in the comet assay in two patients after intoxication with 2,3,7,8-tetrachlorodibenzo-*p*-dioxin. Int Arch Occup Environ Health 77(5):301–306.

Van den Heuval MR, Ellis RJ, Tremblay LA, Stuthridge TR (2002) Exposure of reproductively maturing rainbow trout to a New Zealand pulp and paper mill. Effluent Ecotoxicol Environ Saf 51:65–75.

Van der Kraak GJ, Munkittrick KR, McMaster ME, Portt CB, Chang J (1992) Exposure to bleached kraft mill effluent disrupts the pituitary–gonadal axis of white sucker at multiple sites. Toxicol Appl Pharmacol 115:224–233.

Vonier PM, Crain DA, McLachlan JA, Guillette LJ Jr, Arnold SF (1996) Interaction of environmental chemicals with the estrogen and progesterone receptors from the oviduct of the American alligator. Environ Health Perspect 104(12):1318–1322.

Wallace RA, Selman K (1981) Cellular and dynamic aspects of oocyte growth in teleosts. Am Zool 21:325–343.

Yu Z, Mohn WW (1999) Isolation and characterization of thermophilic bacteria capable of degrading dehydroabietic acid. Can J Microbiol 45:513–519.

Yu Z, Mohn WW (2002) Bioaugmentation with the resin acid-degrading bacterium *Zoogloea resiniphila* DhA-35 to counteract pH stress in an aerated lagoon treating pulp and paper mill effluent. Water Res 36:2793–2801.

Manuscript received October 12; accepted December 18, 2004.

Rev Environ Contam Toxicol 185:93–139

Human Exposure to Lead in Chile

Andrei N. Tchernitchin, Nina Lapin, Lucía Molina,
Gustavo Molina, Nikolai A. Tchernitchin,
Carlos Acevedo, and Pilar Alonso

Contents

Communicated by Lilia Albert J.

A.N. Tchernitchin (✉)
P.O. Box Casilla 21104, Correo 21, Santiago, Chile.

A.N. Tchernitchin · N. Lapin · N.A. Tchernitchin
Laboratory of Experimental Endocrinology and Environmental Pathology, Institute of Biomedical Sciences, University of Chile Medical School, Santiago, Chile.

L. Molina
Institute of Public Health, Ministry of Health, Chile.

G. Molina
Labor Inspectorate, Work Conditions Unit, Ministry of Labor, Chile.

C. Acevedo
University of Los Andes Medical School, Santiago, Chile.

P. Alonso
Regional Public Health Service of Ñuble, Ministry of Health, Chile.

I. Introduction

Lead has been used in human civilizations for about 5000 years and currently has many uses in modern technology. Its uses and toxic properties at high doses have been recognized since antiquity (Waldron 1973). Since the beginnings of lead use, evidence of lead poisoning has been found by medical historians, with dramatic effects on the destiny of ancient civilizations. The fall of the Roman Empire was related to the wide use of lead in paints, water distribution pipes, and wine storage vessels. It was suggested that the declining birthrate, the epidemics of stillbirths and miscarriages, and apparently the increased incidence of psychosis in Rome's ruling class, which may have been at the root of the Empire's dissolution, were a result of exposure to lead in food and wine (Gilfillan 1965). It has been recently proposed that similar or additional effects of human population exposure to lead may contribute to the decline of current societies through lead induced impairment of intelligence (Needleman et al. 1979; Banks et al. 1997), increased tendency to addictions to drug abuse (Tchernitchin and Tchernitchin 1992; Tchernitchin et al. 1999), increase in delinquent behavior (Needleman et al. 1996), or psychological changes such as behavioral difficulties at school (Byers and Lord 1943; Banks et al. 1997).

Acute exposure to high levels of lead causes serious diseases such as lead encephalopathy, which includes brain swelling and can evolve to coma and death. This is not a frequent situation, but may occur in children accidentally exposed to high lead levels, as well as in occupational accidents and attempted suicides.

Chronic exposure to lower levels of lead, which usually does not cause acute symptoms or signs, is a frequent situation affecting urban populations, occupationally exposed workers, and people living in the vicinity of polluting sources. In adults, chronic exposure to lead causes progressive damage to the central and peripheral nervous systems (Needleman et al. 1979; Banks et al. 1997), a moderate increase in blood pressure (Staessen et al. 1994), and effects on both male and female reproductive systems in humans (Winder 1993) as well as in experimental animals (Ronis et al. 1996; Tchernitchin et al. 1998, 2003), causing mainly infertility and an increased abortion rate. It also affects the hematopoietic system (Grandjean et al. 1989; Pagliuca et al. 1990; Graziano et al. 1991); for review of lead interference with the biosynthesis of heme, see NAS-NRC (1972). It depresses thyroid function (Tuppurainen et al. 1988) and causes nephropathy (Weeden et al. 1975, 1979; Ong et al. 1987; Cooper 1988; Cardenas et al. 1993), intestinal colic, gastrointestinal symptoms (Baker et al. 1971; Pagliuca et al. 1990), and damage to the immune system, as shown in humans and experimental animals (Ewers et al. 1982; Jaremin 1983; Cohen et al. 1989; Koller 1990; Lang et al. 1993; Tchernitchin et al. 1997; Villagra et al. 1997). It may also cause effects on chromosomes (Schmid et al. 1972; Deknudt et al. 1973; Al-Hakkak et al. 1986), increase mortality rate (Cooper 1988), and decrease life expectancy.

Children, compared to adults, are more vulnerable to lead in several respects. They typically engage in hand-to-mouth activities (sucking fingers, placing various objects in the mouth), which results in greater ingestion of lead than adults; they present a greater absorption and retention of lead; the exposures occur during sensitive periods of development; and children are generally more sensitive than adults to the toxicological effects of lead at a given blood level (Davis and Grant 1992).

In children, chronic exposure to lead causes a slowing of growth (Schwartz et al. 1986), hearing impairment (Schwartz and Otto 1991; Dietrich et al. 1992), neurobehavioral alterations (Byers and Lord 1943; Needleman et al. 1996; Banks et al. 1997), decrease in the intelligence coefficient (Needleman et al. 1979; Banks et al. 1997), anemia (Schwartz et al. 1990), decreased 1,25-dihydroxyvitamin D_3 plasma levels (Mahaffey et al. 1982) and action (Long and Rosen 1994), slowing of myelin fiber nerve impulses (Davis and Svendsgaard 1990), intestinal colic and associated gastrointestinal symptoms (USEPA 1986), and encephalopathy (USEPA 1986, 1990).

The effects of chronic exposure to lead in adolescents or adults may be reversed, at least in part, after a decrease in blood levels. However, in children, especially at earlier ages, the effects are irreversible and may persist through life, and some effects may be induced by the mechanism of imprinting (*vide infra*).

Lead exposure during pregnancy may lead to congenital anomalies, as reported by Needleman et al. (1984). Lead was found to be associated, in a dose-related fashion, with an increased risk of minor anomalies, mainly hemangiomas and lymphangiomas, hydrocele, minor skin anomalies, and undescended testicles. Lead exposure of pregnant mothers is also associated with low infant body weight at birth (Ward et al. 1987; Bornschein et al. 1989).

Lead exposure during the last months of pregnancy or during early postnatal life causes irreversible changes that persist through life through the mechanism of imprinting. The hypothesis for this mechanism was first proposed by György Csaba, who showed in experimental animals that exposure of fetuses to some hormonally active agents during critical periods of their development induces permanent changes in the action of related hormones (Csaba 1980). These changes can be detected later in adulthood as a modification in the activity of receptors and in the intensity of responses mediated by them (Dobozy et al. 1985; Csaba et al. 1986). This effect of hormones during fetal or neonatal life, permanently modifying the ability of the cells to react to hormone stimulation during adulthood, was named by Csaba "imprinting" (Csaba 1980; Csaba et al. 1986). Work performed in our laboratory demonstrated that, besides hormones, several other agents such as pollutants, including lead, pharmaceuticals, food additives, and normal constituents of food when ingested at abnormal amounts, may activate the mechanism of imprinting and induce persistent biochemical and functional changes that cause the development of diseases later in life (Tchernitchin and Tchernitchin 1992; Tchernitchin et al. 1999).

Exposure to lead during the perinatal period may occur in mothers previously exposed to the metal, presenting increased concentration of lead in bones. Such mothers, during pregnancy or breastfeeding, require additional calcium that is obtained from bone reserves with lead stored together with calcium and is transferred through placenta or breast milk to the fetus or child during the period of life when it is most sensitive to lead. Among the delayed effects induced by this mechanism and evident during childhood, adolescence, or adult life, are neurobehavioral alterations such as decrease in the intelligence coefficient (IQ), aggressive behavior, and a tendency to drug abuse and to delinquent behavior (see Tchernitchin et al. 1999 for the mechanisms involved).

In Chile, besides the classical sources of exposure to lead existing throughout the world (the use of leaded gasoline, lead paints, clusters originated from point sources, food contamination with lead, occupational exposure), there are special cases that deserve additional analysis. Lead exposure has caused considerable damage to Chilean population health. A description of the main sources of exposure, especially those nonfrequent sources affecting large numbers of people, the evolution of Chilean legislation and the role of different organizations and institutions in changing policies, and regulations to prevent exposure to lead, may constitute a useful experience for other countries that are less advanced in prevention and mitigation of lead contamination.

The purpose of this review is to describe available information on the main sources of lead pollution in Chile and report the most conspicuous episodes of lead contamination, describe the most relevant clinical and laboratory findings in exposed population, summarize the evolution of legislation in the country, and analyze proposed actions for remediation and human protection.

II. Primary Sources

Formerly the greater source of lead contamination in Chile was the use of leaded gasoline, mainly in densely populated cities. The progressive decrease in lead content in leaded gasoline, which was banned in April 2001 (SESMA Chile 2002), was the main cause of a decrease in urban lead pollution; however, large amounts of fine lead particles still persist in highly populated cities as city soil and home soil, and as ground contamination near highways, which can be detected in several vegetables such as lettuce grown near highways or within highly populated cities (see following). The Government of Chile issued new regulations for lead in air (DS 136/2000 Chile 2000), allowing 0.5 $\mu g/m^3$ as maximal annual mean.

Another important source of lead contamination was the use of paints with high lead content for painting houses, children's furniture and toys, and other objects. From 1997 on, new legislation regulated lead content in paints (D 374/97 Ministry of Health, Chile 1997); houses, furniture, children's toys, and other objects painted before 1997 are still an important source of exposure to lead, mainly in children. Children's toys, pencils, and other school tools and implements, which children frequently take into their mouth, constitute sources of exposure to lead.

Various degrees of food lead contamination may increase exposure to the metal. Vegetables grown in or near highly populated areas, lead pipes in old dwellings or copper pipes welded with lead, and formerly lead from canned food, especially those cans that had been accidentally deformed, or food stored in opened cans may contribute.

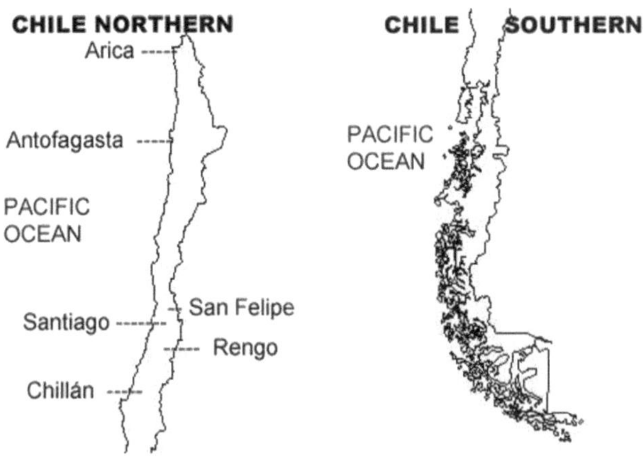

Fig. 1. Map of Chile showing location of cities mentioned in text.

Clusters of different magnitude originated from point sources, of which the most relevant in magnitude was that occurring in Ñuble with the use of wheat flour contaminated with lead, caused by the use of a mill whose stone had been repaired by lead welding. Several other clusters of smaller magnitude are frequently caused by the practice of battery repair and recovery by small enterprises or as family projects, affecting all family members and occasionally those living in the vicinity. Occupational exposure frequently occurs in painters, welders, and mining or smelting workers, among others.

In addition to the traditional sources of exposure to lead common in other countries, in Chile a few special cases of very important environmental contamination occurred, affecting the population of two important cities. The most dramatic case was the storage of powdered lead mineral concentrates at the ports or railroad terminals within the cities of Arica and Antofagasta (Fig. 1), where it would remain until it could be shipped by sea. Lead from this source is the most relevant source of exposure in the cities of Arica and Antofagasta. The second case of lead pollution originated from toxic wastes imported by Promel from the Swedish company Boliden Metal as "toxic raw material for industrial purposes." Local health authorities (1983–1985), although they knew the exact composition of the toxic wastes, informed that "the material is not toxic, anybody can manage it, it cannot be ingested." This material turned out to be hazardous toxic wastes sent to Chile by Boliden Metal and stored without any protection in Arica suburbs between 1983 and 1985, where, a few years later, new dwellings were constructed and residents were exposed for more than 10 years.

III. Lead in Household Paints

Before 1996 there was no regulation on lead content in household paints. Considering that it is an important cause of increased blood levels in children, caused by pica ingestion (mainly leaded paint chips and powder), many other countries have strict regulations. Following several recommendations to the Government by the Colegio Médico de Chile (Chilean Medical Association) that were not taken into consideration, the Colegio Médico measured lead levels in several commercially available paints. Measurements were made by atomic absorption spectroscopy at the Institute of Public Health of the Chilean Ministry of Health. Table 1 shows the lead content (w/w) in household paints (as dry wt of paint) in most commonly sold brands and types in Chile in 1996 (Tchernitchin and Castro 2002).

As soon as the finding of high lead content in household paints sold in Chile was communicated in a press conference, the Ministry of Health organized a commission with the participation of the Colegio Médico to study and propose new standards for lead content in paints. These new standards were approved in 1997 as Regulation 374 from the Ministry of Health, allowing a maximum of 0.06% (dry wt) lead content in paint (D 374/97 Ministry of Health Chile 1997).

The highest lead content in paints corresponded to the brandname of Sherwin Williams (an American corporation that licensed the formula to a Chilean company) that contained nearly 20% lead in the Sherwin Williams yellow brilliant oil enamel. A similar Sherwin Williams paint purchased in Miami, FL, U.S.A.

Table 1. Lead content (w/w) in household paints (as dry wt of paint) in most commonly sold brand names and types in Chile (1996).

Paint trademark	Type of paint	Color	Lead %, w/w dry wt
Sherwin Williams	Enamel oil	Yellow	19.814
Tricolor	Enamel oil	Red	9.518
Sherwin Williams	Enamel oil	Vermilion	4.935
Tricolor	Enamel oil	Cream	3.537
Soquina	Enamel oil	Black	1.035
Sherwin Williams	Enamel oil	Blue	0.921
Sherwin Williams	Enamel oil	White	0.591
Iris	Enamel oil	Green	0.563
Tricolor	Enamel oil	Brown	0.425
Tricolor	Enamel oil	Blue	0.420
Soquina	Anticorrosive	Black	0.388
Soquina	Enamel oil	Calypso	0.338
Iris	Cover oxide	Red	0.329
Soquina	Anticorrosive	Red	0.150
Sipa	Oil opaque	Grey	0.066
Iris	Latex	Green	0.011
Iris	Latex	Black	0.006
Sherwin Williams	Latex	White	0.003
Sherwin Williams	Latex	Blue	0.002
Sipa	Latex	Blue	0.002
Sipa	Latex	Apricot	0.002

Source: Tchernitchin and Castro (2002).

contained nonmeasurable levels of lead (Tchernitchin 2001; Tchernitchin and Castro 2002). This finding was the root of the proposition of the Colegio Médico de Chile (Chilean Medical Association) through the Sustainable Development Council of Chile at the Johannesburg Summit 2002, according to which countries should agree on treaties forbidding exportation of products banned in the countries of their origin because of their health risk, to importing countries without regulations, based on ethical considerations (Tchernitchin and Villarroel 2002, 2003).

Although paints with high lead content supposedly were not sold in Chile after 1997, in many houses walls are painted with lead paint. When a house, especially indoors, is to be painted again, the old paint must be removed; the paint removal procedure widely used in Chile is done with sandpaper or by scraping, leading to the accumulation of high amounts of fine lead-containing dust that remains for a long time inside recently painted buildings and is a source of exposure mainly for children that bring to their mouth toys or other objects from the floor, thus ingesting lead-containing dust adhered to these objects.

IV. Lead in Gasoline

The Metropolitan Area of Santiago (Figs. 1, 2) presents high levels of atmospheric pollution, which are favored by geographic and climatic adverse conditions for the dispersion of pollutants during fall and winter months, and by an increase in economic activity and growth of the city in population and especially in its area. Santiago concentrates 39% of the total Chilean population; its area is 15,554.5 km^2, about 2.1% of the total surface of the country (excluding the Antarctic Territory). Its urbanized surface is about 520 km^2, with 1,710 km^2 used for agriculture; the remaining surface is mountainous.

The city of Santiago is topographically located in a depression, about 500 to 650 m above sea level, and is surrounded by mountains: the highest range, east from Santiago, is the Andean mountains, those closest to the city bordering 3000 m above sea level and higher mountains toward the border with Argentina; the Coastal Range, west from Santiago, with mountains bordering 2000 m above sea level with a narrow strip connecting the Santiago depression with lower altitudes bordering the Pacific Ocean; north from Santiago, mountains about 1200 m above sea level join the Coastal Range with the Andean mountains, and south from the city there exists a very narrow valley that connects the Santiago lowlands with the central valley that becomes wider to the south.

In addition to topographical conditions increasing pollution in the Santiago depression, adverse meteorological conditions exist that worsen during the fall and winter. These include stable atmospheric conditions, mild winds whose direction reverses during night hours, and the atmospheric inversion layer that lowers in altitude and intensifies during certain meteorological episodes during fall and winter, which increase the concentration of various pollutants in air, including lead in breathable particles ($<2.5\ \mu m$ dia).

Winds in the city area are very weak during fall and winter seasons, except the occasional frontal systems that cross the area during the rainy season (May–July,

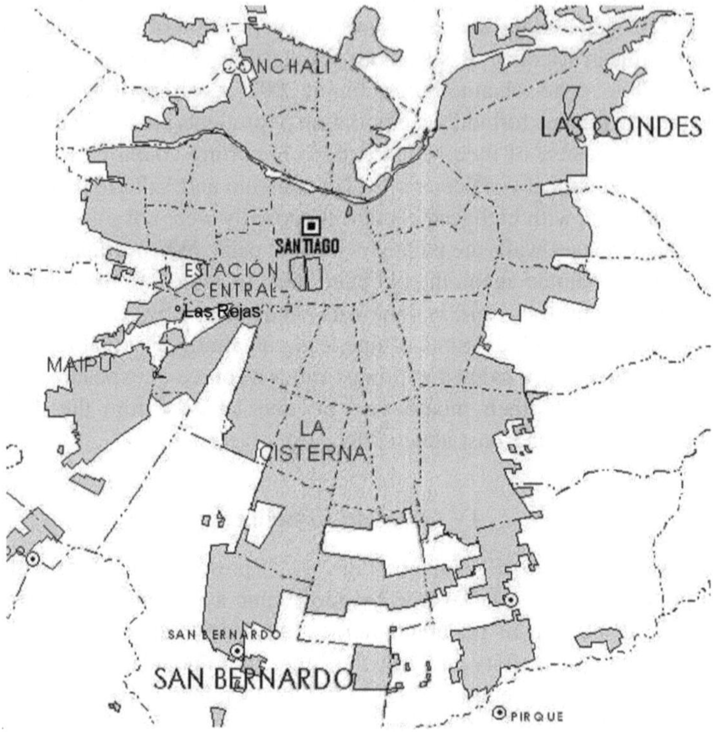

Fig. 2. Santiago Metropolitan Area, indicating the communities or parts of the city mentioned in text. Shaded area corresponds to the densely populated area of Santiago.

2–4 times/mon) during that season. Between the frontal episodes there are very stable atmospheric conditions, and winds are very weak and change their direction from SW during daylight hours to NE during the night. Therefore, the same air may remain for several days or even longer than a week over the city during the fall and winter.

The atmospheric inversion is a layer of warm air that covers a cooler layer beneath and constitutes an obstacle for mixing the layer of polluted city air with the higher clean layers. This inversion layer is thin during the spring and summer and is situated at much higher altitude; therefore, the thickness of this lower layer increases, diluting pollutants with a greater air volume. During the warm season, this inversion frequently breaks up, allowing pollutant dispersion by convection. During fall and winter seasons, this inversion is at low altitude (50–300 m), decreasing the volume of air to dilute city pollutants, and intensifying the inversion (i.e., increasing the difference in temperature between the upper warm layer and the lower cool layer). Most mountains surrounding Santiago are higher than the inversion layer so that the pollutants remain in the depression. The inversion further intensifies under certain meteorological conditions characterized by low wind speeds related to the movement of low atmospheric pressures along the coast, so-called A-type episodes, that intensify the subsidence inversion; i.e., lower air temperature at ground level covered by a higher temperature air layer resulting from heating of air currents descending from the Andean mountains (Gallardo et al. 2000). This situation affects most of Central Chile and in Santiago and other cities gives rise to extreme pollution episodes. These characteristic patterns affecting air pollution in Santiago were described in detail by Ruthllant and Garreaud (1995).

Taking into consideration the above conditions increasing pollution in Santiago metropolitan area, and that one of the components of this pollution is lead, whose main source was the wide use of leaded gasoline, it was necessary to determine blood lead levels in residents of Santiago and their change through the different seasons. This situation was the justification of investigations of blood lead levels and of biological indicators of exposure to lead in Santiago under different climatic conditions, comparing them to those in other less-populated cities. In this context, the amount of lead released from primary sources was also important information that was needed.

The early studies on lead exposure, done in 1979 (Cisternas and Sáez 1980), measured the concentration of the delta aminolevulinic acid in the urine of children 2–5 yr old in childcare centers in central Santiago (downtown), in the suburban Santiago (Conchalí, Las Condes, La Cisterna, Las Rejas, and San Bernardo; see Fig. 2), and Rengo (south of Santiago; see Fig. 1). Figure 3 shows that the urine concentration of delta aminolevulinic acid, an indicator of increased

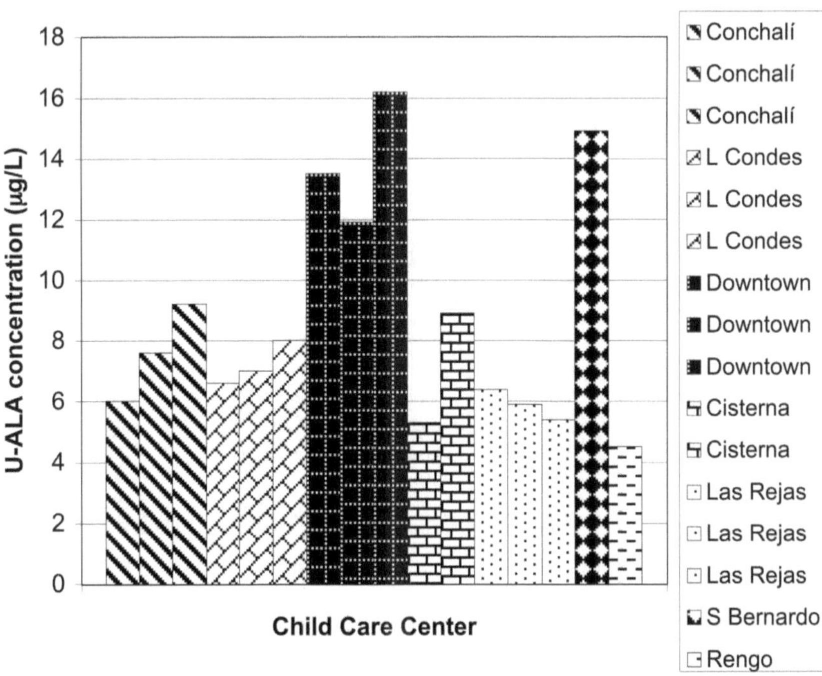

Fig. 3. Urine delta aminolevulinic acid in 2- to 5-yr-old children in 1979, from child care centers in central Santiago, in suburban Santiago, and in Rengo. One to 3 child care centers were investigated in each area. (Modified from Cisternas and Sáez 1980.)

levels of lead in the blood, was much higher in the central part of Santiago than in suburban Santiago and that the lowest levels are in Rengo. Figure 3 calls attention to high levels in San Bernardo, in spite of being suburban, which can be explained by the fact that it is surrounded by industrial areas and is located in the area of influence of Caletones (Gallardo et al. 2000), a copper smelter in the mountains southeast of San Bernardo.

A study performed during 1992 (Ministry of Health, Chile 1993) compared blood lead levels (measured by atomic absorption spectroscopy) from newborn children (umbilical cord blood) and at 6, 12, and 18 mon of age, from Santiago and from the city of San Felipe (see Fig. 1). Figure 4 shows that blood levels were

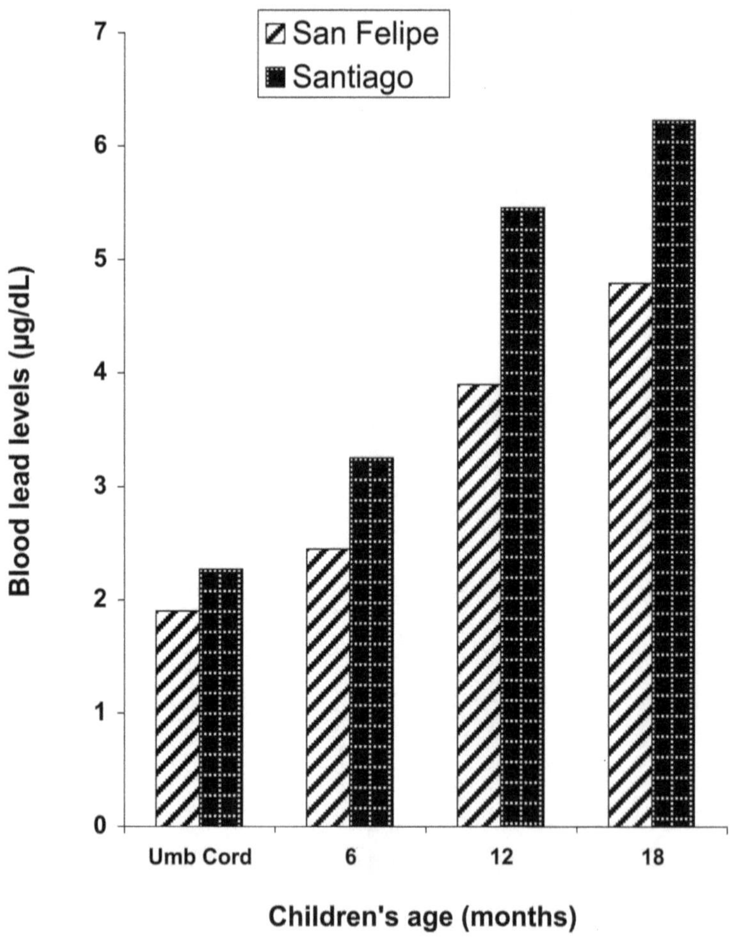

Fig. 4. Lead levels in umbilical cord blood of newborn children and 6-, 12-, and 18-mon-old children from Santiago and San Felipe, 1992 cohorts. (From Ministry of Health, Chile 1993.)

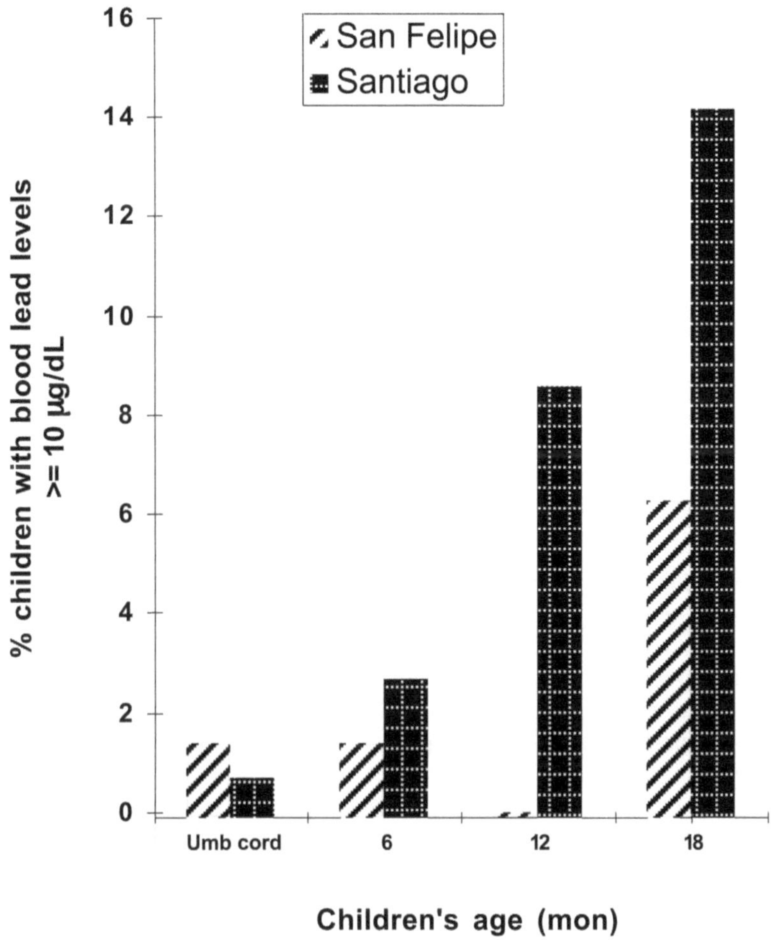

Fig. 5. Percentage of children with blood lead levels ≥10 μg/dL, from Santiago and in San Felipe, 1992 cohorts. (From Ministry of Health, Chile 1993.)

higher in children from Santiago than those from San Felipe and that blood levels increased from birth up to 18 mon of age. Figure 5 shows that the percentage of children with blood levels ≥10 μg/dL, considered at risk, are higher in children from Santiago than in San Felipe.

Comparison of data from Fig. 6 shows the seasonal variation of blood lead levels in children from Santiago and San Felipe (1992–1993). Data from Fig. 7 show the monthly averages of lead in the air of these locations. About a 1-mon delay can be observed between the increase in lead in the air and the increase in blood levels. Both figures also show that lead levels are lower in the city of San Felipe than in Santiago.

Further, a 1992–1995 comparison of children from downtown Santiago, Maipú (a suburb west of downtown Santiago) and the city of San Felipe (Frenz et al.

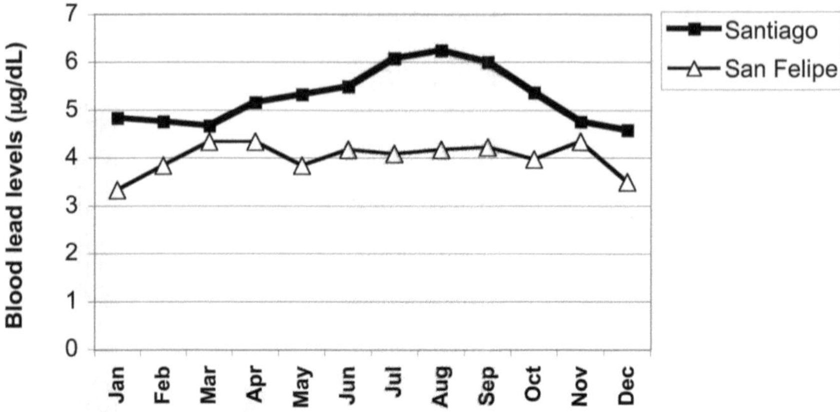

Fig. 6. Blood lead levels in children from birth to the age of 24 mon from Santiago and San Felipe, 1992–1993. (Modified from Frenz et al. 1997.)

1997) (see Figs. 1, 2) showed that blood levels in Santiago were slightly higher than in the remaining locations with San Felipe displaying the lowest levels. Children from all three locations displayed an increase in blood levels from birth to the age of 1 yr; thereafter, lead levels tended to stabilize or slightly decrease (Fig. 8). Figure 9 shows the important difference between Santiago and the remaining locations when comparing percentage of children with blood levels >10 μg/dL.

Four important actions taken by the Chilean government decreased lead concentration in air and lead blood levels in the population: (1) the introduction of

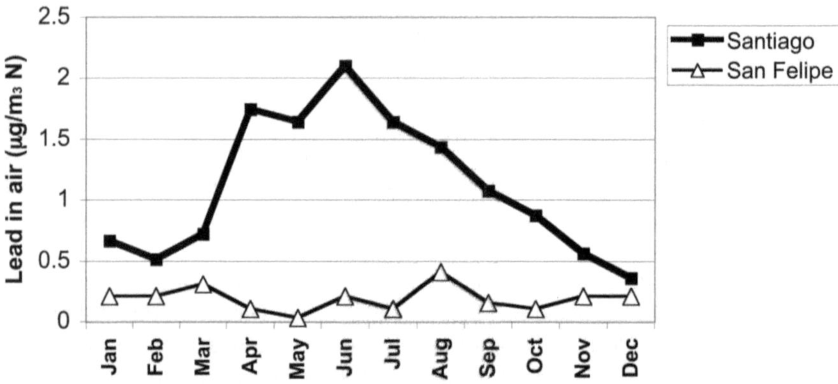

Fig. 7. Lead concentrations in air (monthly averages) in Santiago and San Felipe, 1992–1993. (Modified from Frenz et al. 1997.)

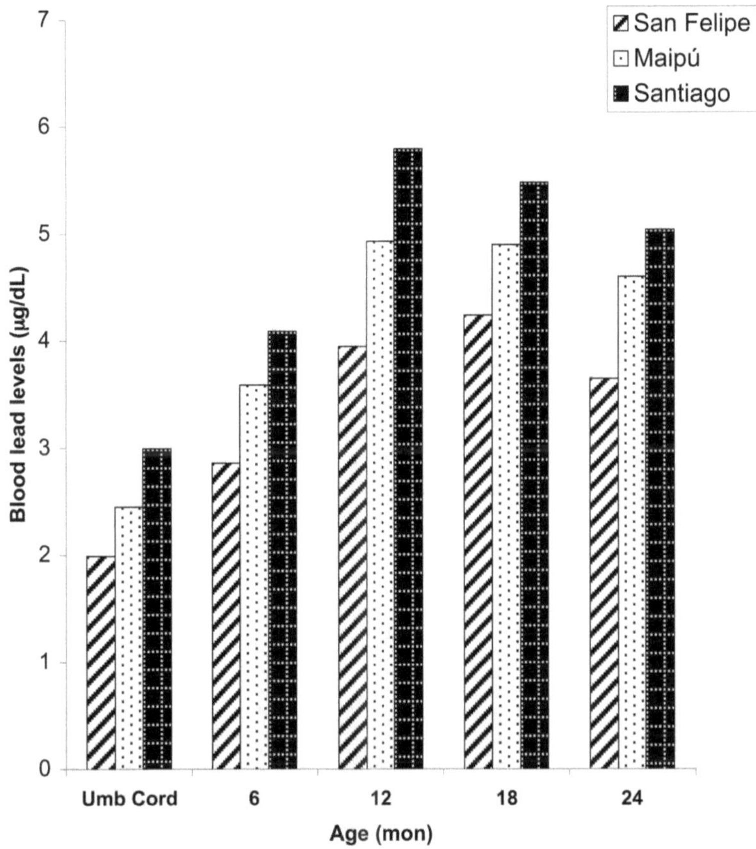

Fig. 8. Effect of age on lead blood levels in 0- to 24-mon-old children from downtown Santiago, Maipú, and San Felipe, 1992–1995. In newborns, blood taken from umbilical cord. (From Frenz et al. 1997.)

unleaded gasoline in addition to leaded gasoline in 1990, which according to the 2003 Prevention and Descontamination Plan for the Metropolitan Area of Santiago approved by Regulation (DS 058/03 Chile 2003) is allowed to have a maximal lead content of 0.013 g/L; (2) the compulsory use of cars with catalytic converters from 1992 models on (which does not allow the use of leaded gasoline because its lead damages the converter); (3) the progressive decrease in lead content in leaded gasoline, which was not available after April 2001 (SESMA Chile 2002); and (4) the approval of the new standard for lead in air (0.5 $\mu g/m^3$ as annual mean level), legally valid from February 2001 (DS 136/00 Chile 2000). Additionally, new regulations on lead content in household paints (see below) and more strict measures for industry emissions of particulate matter contributed to the decrease in lead levels in air and its effects on the population of Santiago.

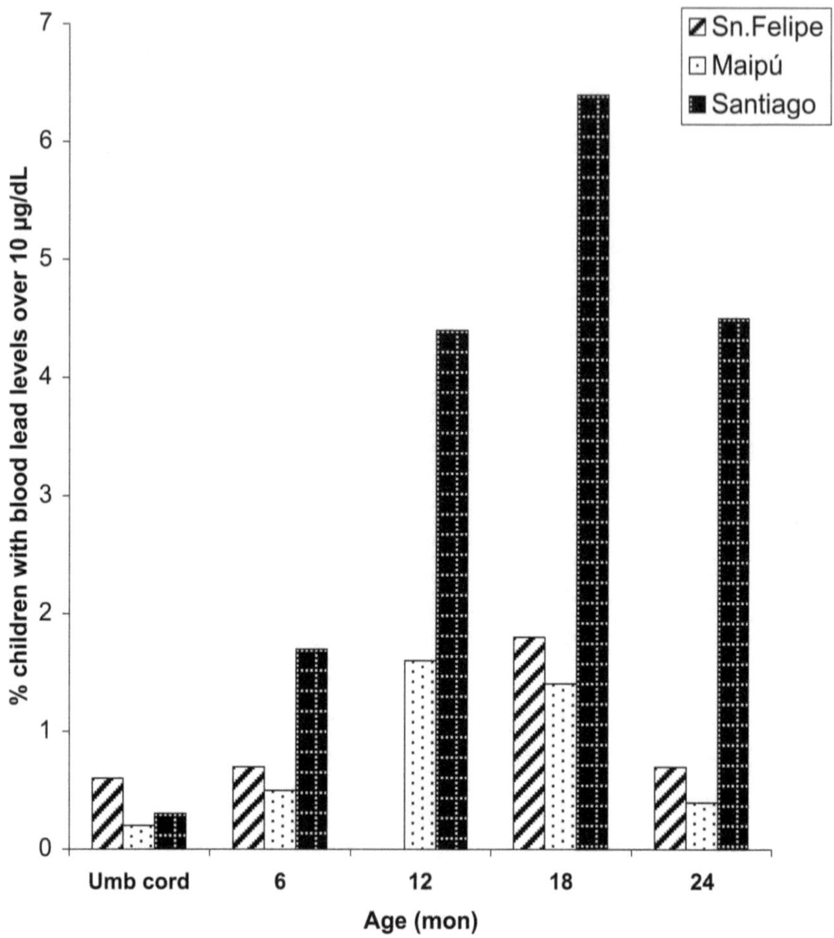

Fig. 9. Effect of age on the proportion of subjects with lead blood levels $> 10\ \mu$g/dL in 0- to 24-mon-old children from downtown Santiago, Maipú, and San Felipe, 1992–1995 (From Frenz et al. 1997.)

According to information provided by the Enviromental Health Service of the Metropolitan Area of Santiago (SESMA Chile 2001), the annual use of leaded and unleaded gasoline in the metropolitan area of Santiago since 1992 is shown in Fig. 10, the concentrations of lead in gasoline in 1997–1999 are shown in Table 2, and the contribution of leaded gasoline to the total amount of lead released into the air of the metropolitan area of Santiago in 1997–1999 is shown in Table 3.

It was assumed that the progressive decrease in lead released into the air originating from leaded gasoline in the Santiago metropolitan area should result

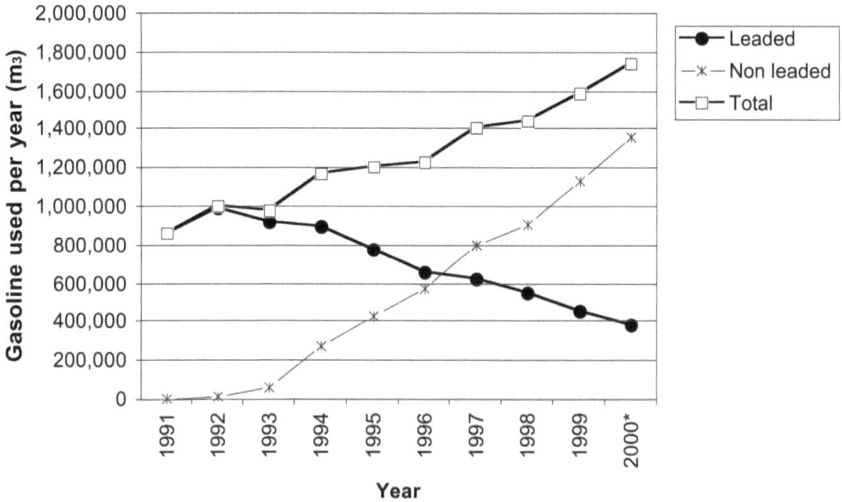

Fig. 10. Annual use of leaded and unleaded gasoline in Metropolitan Area of Santiago, 1991–2000. *, Estimation for 2000. (From SESMA Chile 2001.)

in a progressive decrease in the concentrations of lead in the air of Santiago and, consequently, in the levels of lead in the blood of the general population. A study performed in three locations of the city (La Pintana, Avenida Ossa, and Cerrillos by SESMA Chile (2001) demonstrated the progressive decrease in lead in the air at these locations (Fig. 11). The figure shows the differences between the different locations in Santiago, which can be attributed to the amount of cars using leaded gasoline at these locations and the wind directions at hours of heaviest traffic. Also shown are the seasonal variations in lead concentrations in the air due to climatic conditions, mainly the decrease in the altitude of subsidence inversion and extremely stable atmospheric conditions in fall and winter (*vide supra*).

Table 2. Lead content in leaded gasoline in metropolitan Santiago in 1997–1999.[a]

Year	Pb (g/L)
1997	0.32
1998	0.25
1999	0.19

[a]According to new 2003 regulations, unleaded gasoline maximum allowed lead is 0.013 g/L (DS 058/03 Chile 2003).

Source: SESMA Chile (2001).

Table 3. Lead released through use of leaded gasoline in metropolitan Santiago in 1997–1999.

Year	Lead t/yr	Lead reduction from 1997 (%)
1997	197.1	—
1998	116	41.10
1999	73.25	62.80

Source: SESMA Chile (2001).

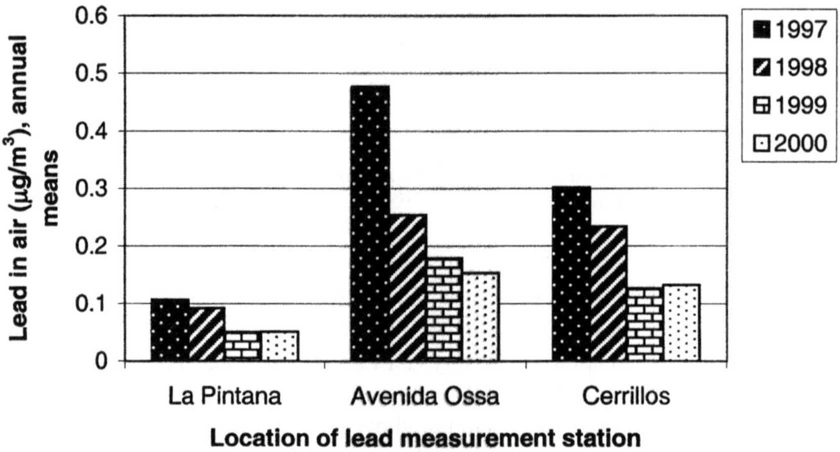

Fig. 11. Annual averages of air lead concentration in La Pintana, Avenida Ossa, and Cerrillos, (1997–2000). (From SESMA Chile 2001.)

V. Lead Exposure Clusters

Clusters of various magnitude originated from point sources, of which the most relevant occurred in Ñuble as a result wheat flour contaminated with lead, caused by the use of a millstone repaired by lead welding. Several other clusters of smaller magnitude are frequently caused by the common practice of battery repair and recovery by small enterprises or as family projects, affecting all the family and occasionally those living in the vicinity.

A. The Ñuble Case: Lead in Wheat Flour

This case of a massive intoxication with lead was reported *in extenso* by Alonso et al. (1997). The massive exposure was discovered with the case of two children that were attended by the Ñuble health service on January 26, 1996, with signs of malnutrition and abdominal and leg pain (Alonso et al. 1997). Both children were interned at the hospital with a diagnosis of malnutrition and serious anemia and a suspicion of maltreatment and family violence. The latter suspicion was disregarded because the parents had gone to different health centers several times during the past 3 mon because of the children's symptoms. It appeared that the head of the family had also been hospitalized recently with a diagnosis of possible gastric ulcer, presenting also with anemia and muscular pain. An hypothesis was drawn that an environmental condition was causing the family's disease. A site visit was coordinated for the next day. Following the suspicion of lead intoxication, a hemogram was taken and revealed punctuate basophilia of erythrocytes in both children, a characteristic sign of lead exposure. The diagnosis of lead toxicity was finally confirmed by atomic absorption spectrometry of blood. The 6-yr-old girl had 104 μg/dL and her 9-yr-old brother 91 μg/dL. The girl died shortly thereafter, despite the initiation of chelation treatment, and the boy developed serious neurological complications.

A professional group from the Regional Public Health Service of Ñuble made the first site visit to the home of the affected children. Sample analyses obtained from the site visit are shown in Table 4. Following the girl's death, several site visits were performed at the family residence, homes in the neighborhood, and the residence of patients frequently attending health centers in the region; these were aimed to determine the causes and source of poisoning and the number of affected individuals. Samples of drinking water, children's toys, house wall paint, food (flour and wheat), soil, and blood were investigated for their lead content at the Institute of Public Health of Chile. Results from this study are shown in Tables 5 and 6.

This study identified the source of lead contamination, which was wheat flour produced in a mill that had a grindstone repaired with lead welding in October 1995. Thereafter, the study included homes in the neighborhood and the residences of patients frequently attending health centers in the region. All these sampled families displayed lead blood levels between 17 and 99 μg/dL. All except the case family worked for the same employer, who gave them wheat flour as part of their salaries, at the San Luis Estate. At this location 60 family heads were employed, and

Table 4. Lead concentrations in samples collected in homes of affected children.

Sample	Lead (μg/g)
Doll dress	0.74
Doll hear	0.01
Plastic spoon	0.01
Paint scraping	7.32
Leaves from a garden plant	23.46
Soil of room for food storage	12.50
Soil surrounding home	37.50
Soil from behind residence	15.00
Well water	Within legal limits (<0.05)

Source: Alonso et al. (1997).

Table 5. Blood lead levels of affected children, their family, and animals living with them.

Family of lead-intoxicated children	Lead, μg/dL	Animals living with the family	Lead, μg/dL
6-yr-old girl (deceased as result of lead poisoning)	104	Pig	25
9-yr-old brother	91	Horse	(nd)
Younger sister	114		
Father	148		
Mother	87		
Grandfather (living elsewhere, frequently visiting the family)	28		

(nd), not detected.
Source: Alonso et al. (1997).

their blood levels were determined. The wheat flour from this estate was milled in three mills; only one of them produced lead-contaminated flour (see Table 6). Wheat samples did not display lead content. At that moment it was found that one of the two millstones had been fractured in several fragments that were welded with lead welding and partially covered by melted lead to join them to a metal arm. As a consequence, 8700 kg of contaminated flour was destroyed.

The exposed population was considered to be 95 families living in the neighborhood and all relatives of people that were noted in a register of clients that gave wheat to be milled in the repaired mill after the repair date, October 23 and 24, 1995, 3 mon before the detection of the first case. A questionnaire was designed for the population to determine consumption of products from the repaired flour mill; it contained questions related to symptoms and signs of lead exposure and about the

Table 6. Lead levels in wheat and wheat flour from various sources.

Wheat and wheat flour from different sources	Lead mg/kg
Wheat flour used by the case family	110
Wheat flour used in Estate San Luis	174.9
Wheat flour used in Estate Serrano	207.2
Wheat flour used by family "2" (the first family identified with symptoms, without any relation to case family)	627.8
Wheat flour from Mill "Quiriquina A"	<1
Wheat flour from Mill "Quiriquina B"	<1
Wheat flour from Mill "Quiriquina C"	112.4
Wheat from other sources	(nd)

(nd), not detected.
Source: Alonso et al. (1997).

Table 7. Subjects presenting symptoms within the total exposed population, according to age.

Age (yr)	*n* (lead-exposed population)	Percent with symptoms	Percent without symptoms
< 6	231	10.9	89.1
6–15	488	10.7	89.3
> 15	1359	22.4	77.6

Source: Alonso et al. (1997).

origin of their flour and the ingested amounts. The Chilean Ministry of Health sent to Chillán, the main city of the Ñuble province, professionals from the Department of Epidemiology of the Ministry, from the Pan-American Health Organization (PAHO), and from the local health services, who applied the following procedure: measurement of zinc proto porphyrine (ZPP) concentrations in blood obtained from digital puncture and a request that the same subjects complete a questionnaire. Four hundred ninety-eight families, totaling 2056 subjects, were studied. Blood collected by venipuncture was obtained for lead content evaluation by atomic spectroscopy from children under 15 years old displaying ZPP values of 3.5 or higher, and from subjects older than 15 years displaying ZPP values of 8 or higher. At the same time, all flour these families had stored at their homes was destroyed (3400 kg). The treatment consisted of blood lead chelation with calcium EDTA (ethylenediaminetetraacetic acid, calcium disodium edetate) at 1000 $mg/m^2/d$ in two daily doses, for 5 d, in hospitalized patients.

The age distribution of those 2056 investigated subjects was 11.24% of children under 6 years of age, 22.68% between 6 and 15 years, and 66.08% older than 15 years. Of them, 18.9% (388 subjects) presented symptoms; these were classified as mainly presenting abdominal symptoms (33.6%), musculoskeletal pain and leg fatigue (28.7%), and other symptoms such as headache, weight decrease, and paleness (37.7%). The proportion of subjects presenting symptoms from the total exposed population, at the different age groups, is shown in Table 7. The lead line in the gums was found in 8 subjects only, all adults.

The results of the 766 persons screened for blood ZPP, and their distribution according to age, are shown in Table 8. According to screening results, 188

Table 8. Positive lead results from screening of zinc protoporphyrine (ZPP) concentrations in blood, according to age.

Age (yr)	*n* (total screenings in lead-exposed population)	*n* subjects displaying positive screening (see text)	Percent positive (see text)
< 6	77	49	63.6
6–15	193	135	69.9
>15	496	246	49.6

Source: Alonso et al. (1997).

Table 9. Subject distribution according to blood lead levels in two age groups.

	Age $\leq$ 15 yr		Age >15 yr	
Blood lead levels (μg/dL)	*n*	%	*n*	%
< 20	4	2.13	0	0
20–40	64	34.04	4	11.76
40–60	102	54.26	18	52.94
> 60	18	9.57	12	35.29
Total	188	100	34	100

Source: Alonso et al. (1997).

determinations of blood lead were performed in the group $\leq$15 years of age and 134 determinations in adults. Table 9 shows the distribution of the subjects according to their age group and lead blood levels. In the children, 137 displayed levels $\geq$40 μg/dL and were treated. Their age distribution is shown in Table 10.

Children were most affected because of the risk of sequelae in this age group. According to the Ñuble study (Alonso et al. 1997), 604 children made up the affected population, which corresponds to 39.9% of the population at risk. Of them, 590 were of school age; therefore, several neurological and behavioral sequelae probably affected this exposed population.

B. Other Clusters

Several other clusters of smaller magnitude are frequently caused by the widespread practice of battery repair and recovery by small enterprises or as a familial productive activity, affecting all family members and occasionally those living in the vicinity.

One typical case occurred in 1995 in Freire, 700 km south of Santiago. There was a court sentence to relocate a lead processing plant due to contamination of soils used for agriculture and that affected the health of residents in the vicinity of the plant for almost 20 years. This sentence was not obeyed, and the activity that was to be finished more than 1 mon ago continued. This motivated the police to close the buildings and seal the furnace where old lead batteries were melted. The

Table 10. Age distribution of children treated with blood lead chelation.

Age (yr)	*n* children treated	Percent children treated
< 1	1	0.7
1–5	20	14.6
6–14	113	82.5
15	3	2.2
Total	137	100

Source: Alonso et al. (1997).

Araucanía Regional Health Service detected, in December 1994, "very high lead levels in the blood" in six children and eight adults in relation to this cluster.

VI. Special Cases

Two special cases of important lead exposures in Chile are analyzed separately: the Antofagasta case of lead contamination that originated from powdered lead mineral concentrate transport and storage, and the Arica case of lead contamination, which originated from both powdered lead mineral concentrate transport and storage and from lead- and arsenic-containing toxic wastes brought from Europe.

A. The Antofagasta Case: Powdered Lead Mineral Concentrates from Bolivia

Antofagasta is a coastal city and port in the north of Chile. The main source of lead pollution was powdered lead mineral concentrates, mainly lead sulfide containing about 26.5% lead according to measurements made at the Institute of Public Health of Chile, Chilean Ministry of Public Health. This mineral, transported through the city and openly stored near densely populated parts of the city, comes by railroad from Bolivia. According to an old international treaty between both countries, Chile is obliged to transport and keep the mineral in its ports until exported by sea to other countries. Until 2002–2003 it was stored in open sites at the Antofagasta port until it could be transported overseas, or at the railroad terminal in Antofagasta, where it usually was stored for years and then transported by trucks to the port, spilling the mineral on its way through the city. The mineral storage areas were not protected against wind; therefore, a significant amount of lead-containing particles was swept by wind to the vicinity, where there were numerous dwellings, the residences of numerous Antofagasta habitants.

The Colegio Médico Regional Antofagasta (Antofagasta Regional Medical Association) drew the conclusion that there was massive exposure to lead in Antofagasta in 1996–1997, based on signs and symptoms suggesting lead toxicity in patients living in the vicinity of lead mineral storage sites. This was the root of the Antofagasta Study by the Colegio Médico de Chile (National Chilean Medical Association) and associated researchers (Andrei N. Tchernitchin, Nina Lapin and Carlos Acevedo) reported below.

The Antofagasta Colegio Médico study included a group of 84 residents living closer than 200 m from the mineral storage place in the Chilean-Bolivian railroad terminal station, who voluntarily agreed to undergo the study. The age distribution of the sample is shown in Table 11.

Blood samples were taken from these residents, who were informed of the purpose of the study and accepted the procedure. Lead levels were measured and an anamnesis and physical examination was performed by one of us (C.A.), searching for neurological signs and symptoms including learning deficit and behavioral alterations. Figure 12 shows mean blood levels for each age group; the ages most affected with the highest levels are 1–4 and 7–10 yr. In the distribution

Table 11. Age distribution in 1997 Antofagasta lead study subjects comprised of residents living less than 200 m from mineral storage site in the Chilean–Bolivian railroad terminal station.

Age (yr)	*n*
1–2	7
3–4	5
5–6	7
7–8	10
9–10	10
11–12	15
13–88	30

of children (1–10 yr old), according to blood levels, only 8% of children 1–10 yr old displayed levels below 10 μg/dL and 8% had levels between 40 and 49 μg/dL. Figure 13 shows the distribution of all ages according to blood levels.

Neurological evaluation of the study group of these 84 residents (all ages) revealed that the percentage of subjects displaying neurological manifestations increased in groups with higher blood lead levels (Fig. 14). A similar situation was observed for the children 1–10 yr old (Fig. 15).

Neurological manifestations were evaluated as the sums of scores of the different neurological signs and symptoms from anamnesis and physical examination by the neurologist from our research team (C.A.), which included learning deficit and behavioral alterations. For each symptom or sign, value 1 was assigned to moderate intensity and 2 for severe. Results are shown for all age groups (Fig. 16)

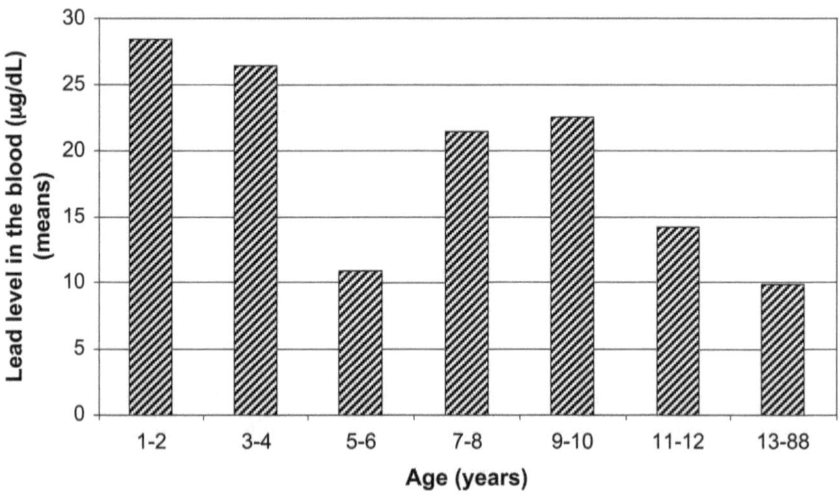

Fig. 12. Blood lead levels in subjects living less than 200 m from lead mineral storage site in Antofagasta railroad station (1997).

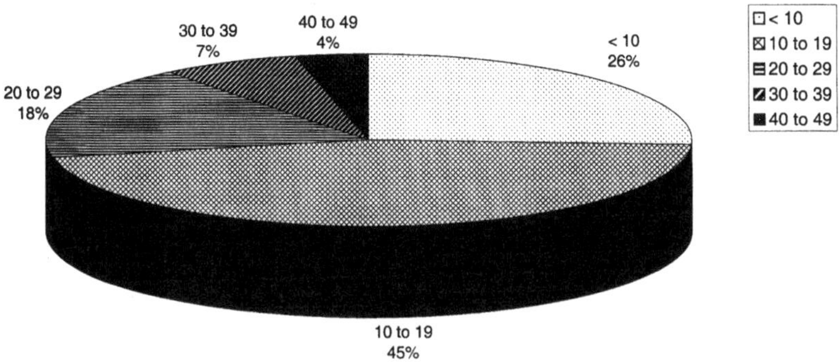

Fig. 13. Blood lead levels (μg/dL) distribution in the group of residents (all ages) living less than 200 m from lead mineral storage site in Antofagasta railroad station (1997).

and for the children 1–10 yr in age (Fig. 17), where the intensity of neurological manifestations is correlated with blood lead levels in both the children and all age groups.

Other studies continued with the investigation of blood levels in Antofagasta residents, classifying them into three groups according to the place of their residence: (1) in the vicinity of the lead mineral storage place at the railroad station, (2) in the vicinity of the port, where the concentrated mineral also was stored, and (3) a control group situated far away from contamination sources and the

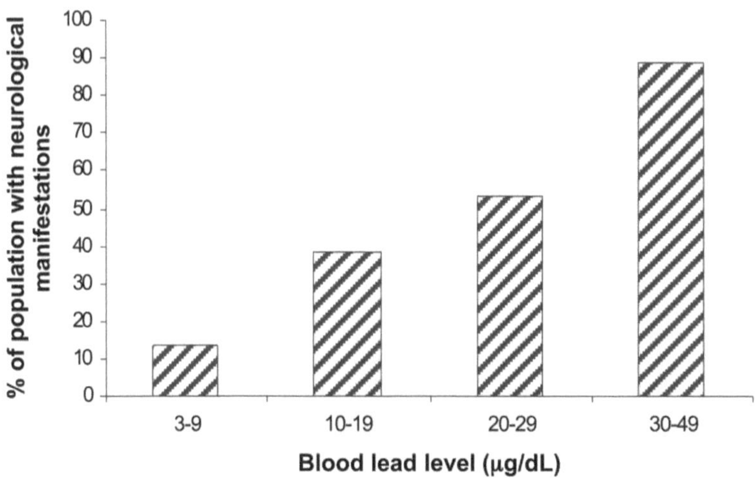

Fig. 14. Percentage of subjects displaying neurological manifestations, according to blood lead level, in the group of residents (all ages) living less than 200 m from lead mineral storage site in Antofagasta railroad station.

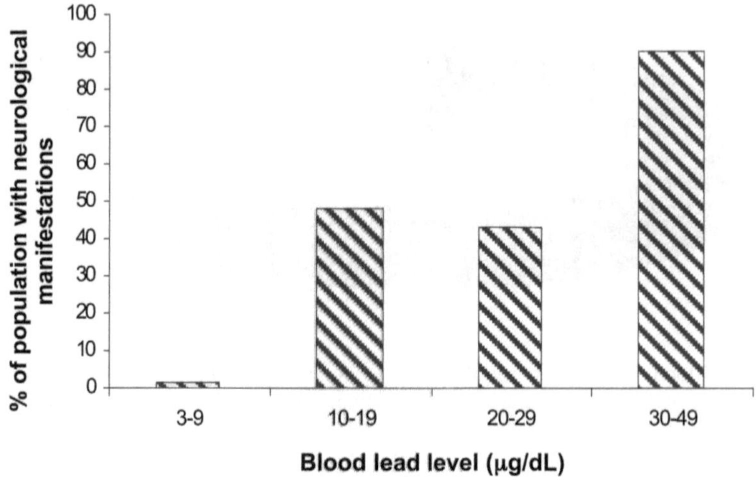

Fig. 15. Percentage of subjects displaying neurological manifestations, according to blood lead levels, in the group of yr 1- to 10-yr-old children living less than 200 m from lead mineral storage site in Antofagasta railroad station.

predominant wind direction from any of the contaminated places in Antofagasta. These sites and their residents were evaluated for lead concentrations in air, soil, and drinking water and lead concentrations in the blood during 1998 (Sepúlveda et al. 2000). Table 12 shows that the geometric means of lead levels in blood were more than doubled in the population living near the railroad station than in the control location: levels of residents near the port had lower levels than in

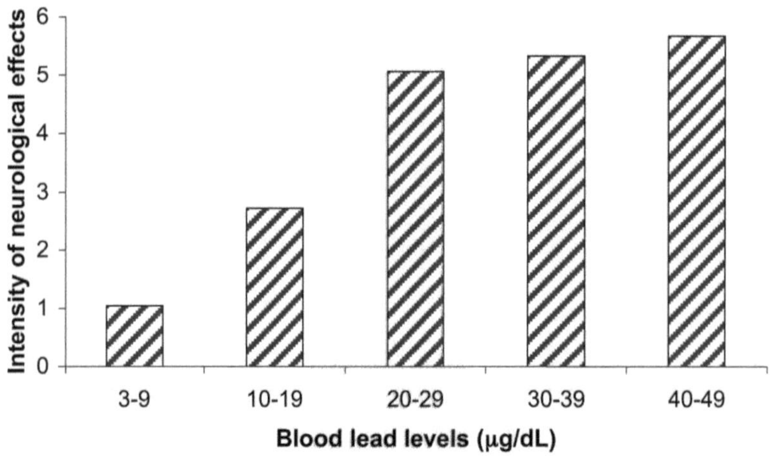

Fig. 16. Intensity of neurological manifestations in the group of residents (all ages) living less than 200 m from lead mineral storage site in Antofagasta railroad station, and their blood lead levels.

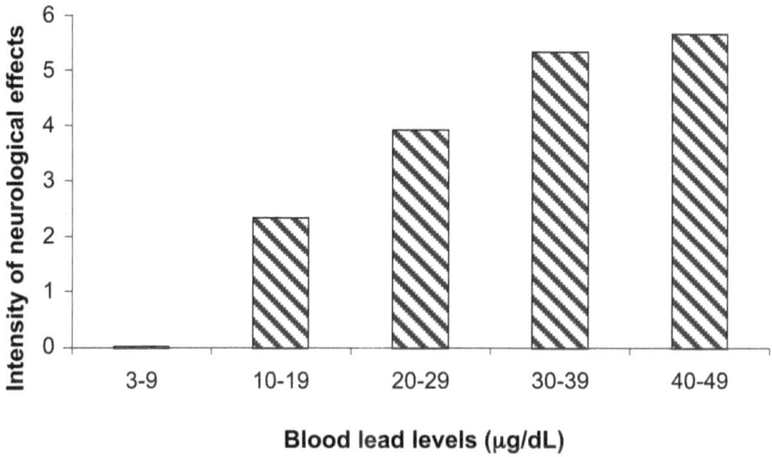

Fig. 17. Intensity of neurological manifestations in the group of residents (1–10 yr old) living less than 200 m from lead mineral storage site in Antofagasta railroad station and their blood lead levels.

those near the railroad station but considerably higher than in the control location. Samples of lead levels in the soil around residences near the railroad, or in the dust over the dwellings or inside homes, had much higher levels than in control locations, some of them reaching >3 g lead/kg; the highest level in the control location was 321 mg/kg. The study did not investigate the location near the port. Levels of lead in respirable particulate matter (<10-μm-diameter particles) in air were two fold higher in the port than near the railroad station. Lead in air seemed to be lower in the control location, but this information cannot be taken into consideration because it comes from measurements made over 2 mon, in contrast to the other measurements that averaged 6 months, and the station-dependent climatic conditions may alter lead concentration in air.

Table 12. Lead concentrations in blood and environmental lead in three areas of Antofagasta, 1998.

	Lead (μg/m^3) in respirable particulate matter (< 10 μm diameter)[a]	Lead content in soil (mg/kg)	Lead in blood (μg/dL), geometric means
Port area	0.28 (6 mon)		6.890
Railroad station area	0.14 (6 mon)	81–3159	8.671
Noncontaminated control area	0.10 (2 mon)	51–321	4.221

[a]Time in parentheses is sampling period over which data were averaged.
Source: Modified from Sepúlveda et al. (2000).

Table 13. Percentage of subjects with more than 10 and 20 μg/dL lead in their blood in three areas of Antofagasta 1998.

	Sample size n	Blood lead $\geq 10\,\mu$g/dL		Blood lead $\geq 20\,\mu$g/dL	
		n	% of sample	n	% of sample
Port area	54	17	31.5%	5	9.3%
Railroad Station area	432	205	47.5%	43	10%
Noncontaminated control area	75	0	0%	0	0%
Total	561	222	39.6%	48	8.6%

Source: Modified from Sepúlveda et al. (2000).

Table 13 shows that all 75 subjects living in the noncontaminated area displayed lower lead levels in blood than in those in both contaminated areas, which had blood lead levels $\geq$10 or 20 μg/dL, the highest being in residents near the railroad station. Table 14 shows the age distribution of subjects with >10 and 20 μg/dL lead in blood in the above three areas of Antofagasta. Table 15 illustrates the association of geographic location of residence and blood levels in exposed locations. Further, the Sepúlveda et al. (2000) study demonstrated that the distance of dwellings from lead storage sites, their geographic location, and their antiquity were significantly associated with high blood levels, and that multivariate models disclosed that people living in contaminated sites (near storage places of the railroad or near the port) have odds ratios of 24.85 and 22.56, respectively, for blood lead levels $\geq$7 μg/dL.

Carboncillo Beach, City of Antofagasta In 1991, a heavy rain event occurred in the city of Antofagasta and its surroundings, which is unusual because of an

Table 14. Age distribution of subjects with more than 10 and 20 μg/dL lead in blood, in three areas of Antofagasta, 1998.

Age (years)	Sample size n	Blood lead $\geq 10\,\mu$g/dL		Blood lead $\geq 20\,\mu$g/dL	
		n	% of sample	n	% of sample
<1	55	18	32.7	3	5.5
1	60	26	43.3	6	10
2	60	31	51.7	11	18.3
3	63	37	58.7	8	12.7
4	72	34	47.2	10	13.9
5	73	34	46.6	4	5.5
6	68	29	42.6	4	5.9
7	33	11	33.3	1	3.0
Total	484	220	45.5	47	9.7

Source: Modified from Sepúlveda et al. (2000).

Table 15. Geographic location of residence and blood lead levels in exposed locations, 1998.

Lead levels in blood	Degree of exposure to lead	Percent of population
Lead $\geq 20\,\mu g/dL$	High	21.5
	Intermediate	10.1
	Low	3.8
Lead $\geq 10\,\mu g/dL$	High	63.6
	Intermediate	45.0
	Low	37.5

Source: Modified from Sepúlveda et al. (2000).

extremely dry climate with an average rainfall of 1.7 mm/yr, 30-yr average. An alluvial flood of water crossed part of the city, carrying a large amount of soil, and swept off stored mineral concentrates from the storage site at the railroad station. This material was carried to the ocean, and presumably is the main source of lead contamination at Carboncillo Beach, one of the city's beaches. The Regional Public Health Service of Antofagasta (2000) reported very high levels of lead in this location. In the beach intertidal sand it reached 2710 mg lead/kg; on the seafloor near the beach, at 2–5 m beneath the surface, it reached 1930 mg lead/kg in sediments, and at 10-m depth it reached 1080 mg/kg in sediments, compared to reference sites in Antofagasta (23–61 mg/kg lead).

We investigated a case of a family who moved to a house located 50 m from the Carboncillo Beach and had been living there for about 2 yr (1999–2001). All three members of the family presented severe neurological problems and irreversible brain damage assessed by single photon emission computed tomography (SPECT) for cerebral perfusion with Tc-99m HMPAO (hexamethyl propyleneamine oxime). The family moved to Santiago, and 6 mon after moving from Antofagasta received the diagnosis of lead exposure sequelae. Blood levels of lead 6 and 12 mon after leaving Antofagasta are shown in Table 16.

In all three family members, SPECT for cerebral perfusion with Tc-99m HMPAO at 6 mon after leaving Antofagasta revealed multifocal hypoperfusion of uneven distribution and several broad areas with hypoperfusion in the brain, and additionally

Table 16. Changes in blood lead levels in family after moving from lead exposure source in Carboncillo Beach, Antofagasta.

Family member	Father	Mother	Son
Age when leaving Antofagasta	38 yr	28 yr	2 yr
Blood lead levels 6 mon later	$34.6\,\mu g/dL$	$36.9\,\mu g/dL$	$32.2\,\mu g/dL$
Blood lead levels 12 mon later	$1.2\,\mu g/dL$	$2.2\,\mu g/dL$	$6.0\,\mu g/dL$

bilateral anterior frontal increased perfusion hyperfrontality. One year after leaving Antofagasta, the hypoperfusion areas worsend as well as the hyperperfusion in the anterior frontal brain, which increased further in the father. The alterations progressed only slightly in the child and showed a moderate regression in the mother. In the father, neurocognitive testing revealed damage as an incomplete Wisconsin test at the 14th percentile. Electromyograms and nerve impulse conduction velocity measurements in the father revealed a distal polyneuropathy, predominantly sensitive, and a psychological examination revealed depression, psychoorganic damage mainly in short-term auditory and visual memory, and possible alteration of personality. In the child, the psychological analysis showed a diagnosis of aggressive behavior at 3 yr of age. These findings reveal that although lead blood levels decreased to low levels after 1 yr of living away from the exposure source, the neurological sequelae persisted and even worsened in two of the three family members.

B. The Arica Case: Toxic Wastes from Sweden and Mineral Concentrates from Bolivia

Arica is the most northern Chilean city and port, situated a few kilometers south of the Chilean border with Peru. There were two main sources of lead pollution: transport and storage of lead mineral concentrates from Bolivia and the "import" of toxic wastes from Europe.

The first source of lead exposure in Arica was from powdered lead mineral concentrates, mainly lead sulfide containing about 26.5% lead. Until recently, this mineral was transported through the city and openly stored near densely populated parts of the city, in open places near the railroad until 1993, and from then on at the port itself. The mineral comes by railroad from Bolivia. According to an old international treaty signed between both countries in 1904, Chile is obliged to allow free transit of Bolivian merchandise through Chilean territory to its ports until it can be exported by ship to other countries. According to available information, 340,000 t of concentrates are shipped every year through Chilean ports in Arica and Antofagasta. The storage areas were not protected against wind; therefore, an important amount of lead-containing particles was blown throughout the vicinity, where there are many homes, whose inhabitants are exposed to high levels of lead in air and soil.

The second source of exposure was to "toxic raw material for industrial purposes" imported by Promel from the Swedish company Boliden Metal. This material turned out to be hazardous toxic wastes sent to Chile by Boliden Metal and deposited without any protection in the suburbs of Arica between 1983 and 1985 where, a few years later, new low-income dwellings were constructed and residents were exposed for more than 10 yr. Although the local health authorities (1983–1985), knew the exact composition of the toxic wastes, they allowed the entry of these wastes to Chile and informed that "the material is not toxic, anybody can manage it, it cannot be ingested."

To the lead pollution just described there is an added adverse situation: Arica has an extremely arid climatic condition. Annual rainfall is about 0.5 mm/yr (30-yr avg) which prevents lead movement from the surface to lower soil layers or its removal by rain as occurs under different climatic conditions.

The Colegio Médico Regional Arica (Arica Regional Medical Association) concluded that there was massive exposure to lead around 1997, based on signs and symptoms suggesting lead and arsenic toxicity in patients living in the vicinity of these toxic wastes. This was the origin of the Arica Study (1998 and 1999) performed by the Colegio Médico de Chile (National Chilean Medical Association) and associated researchers (A.N.T., N.L., and N.A.T.) reported below. This study also discovered a very important lead exposure source, the old lead mineral concentrate storage area that was not suspected previously, and it was included in the study.

Lead Measurements in Blood, 1998 and 1999 We investigated blood levels in exposed and control children from Arica in two stages. The first was in 1998 in Poblacion Los Industriales and Cerro Chuño and in Villa Santa Maria. The second, which included a larger number of subjects, was in 1999 and included the same two areas from Arica and a third additional area, Pampa Nueva, which was used as a control area supposedly with low environmental lead levels. In both studies, blood samples were taken from these residents who accepted the procedure. Levels were measured by the electrochemical method (Leadcare). Additionally, in the second study a computational test to detect a delay in hand-motor response to three sensorial stimuli was applied, and samples of head hair were taken to measure lead content. Most children in the 1998 study were also included in the 1999 studies.

Table 17 describes the general population investigated in two areas of Arica 1998 whose lead levels and other parameters were again investigated in 1999. One of the areas, Poblacion Los Industriales and Cerro Chuño had new low-income dwellings constructed near the disposal area where toxic wastes were stored by Boliden Metal and were still there during the study. Those investigated lived within 200 m of the waste location. The general population from the other area, Villa Santa Maria, was initially intended to be the control subjects, but displayed very high blood levels. It was found later that the Villa was composed of new houses built for a middle- or higher-income population near an old storage site for lead mineral concentrates from Bolivia. Possibly some of the houses were constructed

Table 17. Age distribution in the Arica 1998 lead study group, comprising residents living at two locations, whose blood lead levels were measured again in the 1999 study.

Age (years)	Santa Maria	Los Industriales and Cerro Chuño
5	2	6
6	1	9
7	4	12
8	2	5
Total	9	32

Table 18. Changes in blood lead levels in the same children from 1998 to January 1999 Arica study groups in Santa Maria (S M) and Los Industriales and Cerro Chuño (L I & C C).

Location	Date	n	Mean Pb, μg/dL	Percent $\geq$10 μg/dL	Percent $\geq$20 μg/dL	Percent $\geq$30 μg/dL
S M	1998	9	21.8	100	67	11
S M	Jan. 99	9	20.1	89	44	11
L I & C C	1998	32	8.3	38	0	0
L I & C C	Jan. 99	32	8.4	31	1	1

on the old storage site that never was cleaned up. Table 18 shows blood lead levels in children 5 to less than 9 yr old living in both areas and compares the change of blood levels between the first and second measurements.

The 1999 study comprised a larger number of exposed children than in the 1998 study, mainly from Poblacion Los Industriales and Cerro Chuño, and additionally a new area of Arica, Pampa Nueva, located far from lead sources, was incorporated into the study as a control area. Subjects from Poblacion Los Industriales and Cerro Chuño and those from Pampa Nueva belong to a low-income socioeconomic class whereas residents from Villa Santa Maria are in the higher-income group with higher educational status. Figure 18 shows blood levels in children from these areas, distributed according to age. Blood levels in children from Villa Santa Maria were very high and displayed the highest values in younger ages, but tended to decrease rapidly with age, which does not occur in children from Los Industriales and Cerro Chuño. Levels in children from Pampa Nueva were the lowest, except in one cluster of three subjects of 9 to <10 yr age, who lived on the same street <100 m of each other. In the homes of two, there were some labor activities with lead.

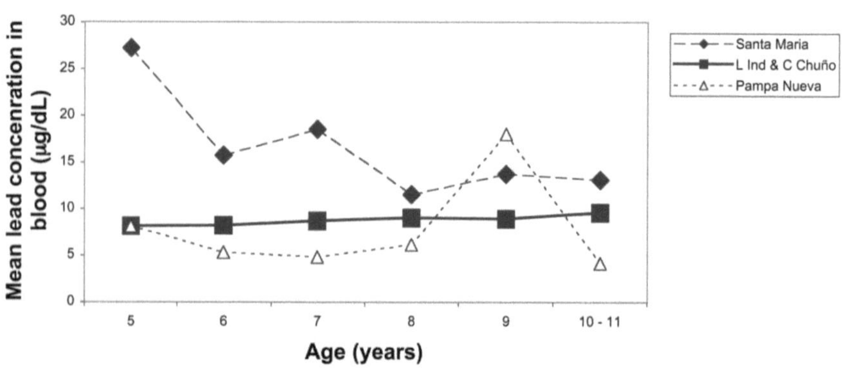

Fig. 18. Blood lead levels according to age in the Arica 1999 lead study universe for children living at three locations whose hair lead levels were measured.

Table 19. Blood lead levels in the group 5- to 8-yr-old children living in three areas of Arica, 1999.

	Santa Maria	Los Industriales & Cerro Chuño	Pampa Nueva
n	12	116	13
Mean Pb, μg/dL	18.1	8.5	5.8
SD	8.5	5.7	2.5

SD: standard deviation.

Considering this cluster of subjects with high blood levels, Tables 19 and 20 show blood levels in children 5 to <9 yr old distributed according to their residence area and the children's distribution according to their blood levels. Tables 21 and 22 show the same results in all children in the study 5 to 11 yr old. In Tables 19 and 20, a clear difference is observed between levels from Villa Santa Maria, the highest levels, and Los Industriales and Cerro Chuño, and the controls residing in Pampa Nueva.

Measurements of Head Hair Lead At the same time blood was taken from children in the three areas of Arica (1999), a sample of head hair was taken for lead measurement. Table 23 shows lead content in children 5 to <9 yr old, and Table 24 shows values in 5 to 11 yr old. The same differences described for lead in blood were observed in hair.

Evaluation of Effects Associated with Lead Taking into consideration the effect of lead on central and peripheral nervous system, specially on easily measurable parameters, such as the decrease in the nerve impulse transport by myelin nerves, which was supposed to cause a delay in stimuli motor response, we designed a computer program (Silva et al. 1999) that allows the quantification of the delay between sensitive stimuli and a motor response to that stimuli. Three different kinds of stimuli were applied: (a) a simple visual stimulus such as a simple object suddenly visible on a computer screen, (b) a more-complex visual stimulus that requires brain processing to discriminate among several visual images appearing

Table 20. Distribution of children 5 to 8 yr old, according to blood lead levels in the groups from three areas of Arica, 1999.

Level (μg/dL)	Santa Maria	Los Industriales & Cerro Chuño	Pampa Nueva
0–9.9	2	81	12
10–19.9	6	31	1
20–29.9	3	1	0
30–39.9	1	3	0

Table 21. Blood lead levels in the group of 5- to 11-yr-old children living in three areas of Arica, 1999.

	Santa Maria	Los Industriales & Cerro Chuño	Pampa Nueva
n	21	170	21
Mean Pb, μg/dL	16.1	8.7	7.2
SD	8.1	6.3	5.2

on the computer screen, and (c) an auditory stimulus (acute sound) produced by the computer. Each kind of stimulus was applied several times each, to allow learning.

Table 25 shows the delay in response to auditory stimulus in children of the same age with blood levels ≥10 μg/dL compared to children with <10 μg/dL. Results show that there is a delay in response to sensory stimulus starting at <10 μg/dL, at much lower levels than those causing a decrease in the velocity of nerve conductance through motor myelinated nerve. It is possible to speculate that in addition to a delay in motor nerve impulse transmission, lead additionally causes some delay in another site, at the central nervous system or at the synapse level. These possibilities deserve future studies to better understand the mechanisms of lead interaction with nerve cells at blood lead levels >10 μg/dL.

Actions Taken The Colegio Médico de Chile informed the local and national health authorities about these results and further informed the local population through the press and through a health campaign. These activities of the Colegio Médico motivated implementation of several measures by the government to prevent or mitigate damage, summarized as follows:

a. Removal of toxic wastes from the Los Industriales and Cerro Chuño area of Arica 2 km away to Quebrada Encantada in 1998–1999, an uninhabited creek, and covered these wastes with 50 cm soil protected by surrounding the site with a 2.20-m-high wall.

Table 22. Distribution of 5- to 11-yr-old children according to their blood lead levels, in the groups from three areas of Arica, 1999.

Level (μg/dL)	Santa Maria	Los Industriales & Cerro Chuño	Pampa Nueva
0–9.9	5	121	17
10–19.9	10	41	3
20–29.9	5	3	1
30–39.9	1	5	0

Table 23. Lead concentration in hair from 5- to 8-yr-old children living in three areas of Arica, 1999.

	Santa María	Los Industriales & Cerro Chuño	Pampa Nueva
n	11	96	8
Mean Pb, μg/g hair	52.5	18.3	16.1
SD	35.6	10.4	8.1

b. After waste was removed from the original site, it was covered with about 60 cm soil, partially paved, and surrounded with walls while waiting to pave that area completely.
c. Soil samples from streets and squares for lead measurements.
d. Cleaning of houses and roofs in Los Industriales and Cerro Chuño.
e. Decontamination of the railroad lead mineral storage area Chinchorro near Villa Santa María.
f. Soil samples from the railroad storage site, surrounding area, and streets in the Villa Santa María.
g. Cleaning of houses and roofs in Villa Santa María.
h. Cleaning of the Arica Port lead mineral storage area, and construction of new storage areas to decrease lead contamination of surrounding area.
i. Measurement of blood lead levels in exposed population.
j. A protocol of actions and/or treatment of children according to blood lead levels, which consisted of family education, Fe supplementation in diet, calcium supplementation in diet, hemogram evaluations, new lead level controls in blood, investigation of home conditions that favored lead exposure, and therapeutic procedures, according to blood levels.

Measurements of Blood Lead Levels by the Local Health Service of Arica The local Health Service reported the evaluation of 155 individuals among children and adults living in Villa Santa María within the last semester of 1998 and 1999 (Servicio de Salud de Arica 2000). Blood was collected by digital puncture for lead measurement by the electrochemical method (Leadcare) or by venupuncture for measurement by atomic absorption spectroscopy at the Institute of Public Health

Table 24. Lead concentration in hair from 5- to 11-yr-old children living in three areas of Arica, 1999.

	Santa María	Los Industriales & Cerro Chuño	Pampa Nueva
n	25	169	23
Mean Pb, μg/g hair	55.7	18.7	20
SD	32.6	11.2	24.9

Table 25. Delay in response to auditory stimulus in children with lead blood levels $\geq 10\,\mu g/dL$.

Blood lead level (μg/dL)	Response (sec) in 8-yr-old children		Statistical difference[a]	Response (sec) in 9-yr-old children		Statistical difference[a]
	< 10	≥ 10		< 10	≥ 10	
n	22	12		35	12	
Mean	0.469	0.522	2 P < 0.001	0.422	0.476	1 P < 0.05
SD	0.075	0.088		0.029	0.070	
SEM	0.016	0.025		0.005	0.020	

SEM: standard error of the mean

[a]Least significant difference (LSD) a posteriori test.

(ISP) of the Ministry of Health. Results of blood levels $\geq 10\,\mu g/dL$ are shown in Table 26.

The Local Health Service of Arica reported (Servicio Municipal de Salud Arica 2001; Servicio de Salud de Arica 2001) that it took 4990 blood samples from subjects from Los Industriales and Cerro Chuño, composed of 3240 children (65%) and 1750 adults (35%), for lead measurement. Results showed 4411 samples were $<10\,\mu g/dL$ and 579 $\geq 10\,\mu g/dL$. Of the "positive" 579 subjects, blood was collected from 538 (92%) and measured at the Public Health Institute (ISP) of the Ministry of Health of Chile. Results showed 131 subjects had $\geq 10\,\mu g/dL$ and 407 $< 10\,\mu g/dL$, of 120 children (<15 yr old) and 11 adults. Data from the 120 children showing age distribution and blood levels are given in Table 27.

Table 26. Subjects from sample population of Villa Santa María displaying high blood lead levels according to evaluation by the Health Service of Arica in 1998–1999.

Level (μg/dL)	Electrochemical method (Leadcare)	Atomic absorption spectroscopy, 1998	Atomic absorption spectroscopy, 1999
Total subjects			
10–19.9	32	51	15
20–29.9	3	16	4
30–39.9	0	2	1
Children under 9 yr of age			
10–19.9	11	28	9
20–29.9	1	9	3
30–39.9	0	1	1

Source: Servicio de Salud de Arica (2000).

Table 27. Age and blood lead level distribution of children from Población Los Industriales and Cerro Chuño displaying blood lead levels $\geq 10\,\mu g/dL$, from a study of the local Health Service during 2000.

Age	10–14 μg/dL	15–19 μg/dL	20–34 μg/dL	Total
< 2 yr	16	7	1	22
2–5 yr	48	12	3	63
6–9 yr	19	2	2	23
10–14 yr	7	3	2	12
Total	90	22	8	120

Source: Servicio Municipal de Salud Arica (2001); Servicio de Salud de Arica (2001).

VII. Other Sources

The most relevant sources of lead, affecting mainly the urban population, were leaded gasoline and high lead content in household paints. New regulations on these sources contributed to an overall decrease in exposure to lead in Chile. Lead originating from gasoline is still present at high levels in soil of large cities and from the vicinity of high-traffic highways. Dwellings painted before 1998 also constitute an important source of lead contamination.

Other sources of lead in Chile are children's toys and school implements, occupational exposure to lead, and lead contamination of food. Several industrial activities occur, mainly in cities. Many of them recover lead from car batteries (some are licensed, others are clandestine), and there are industries melting metals painted with leaded paints. There are many families working at these activities in their homes, such as recovering lead from batteries and making different objects with lead, and some working in stained glass, and welders, etc.

Children's Toys and School Implements Lead exposure may occur from children's toys, pencils, and other school tools and implements, which children frequently take into their mouth. These items may be imported from countries without lead regulations.

Occupational Exposure Occupational exposure frequently occurs in painters, welders, glass handicraft workers, and miners and smelting workers. Atmospheric lead may be originated from industries melting metals painted with leaded paints, affecting their workers and the population living in the vicinity.

Food Contamination Food may be a source of various degrees of exposure to lead. The use of ceramics covered with low-quality leaded paint is not frequent in Chile; however, these ceramics can be imported from countries without regulations. Vegetables growing in or near highly populated areas and specially the metropolitan area of Santiago or near heavy traffic highways concentrate lead from soil, adding to the lead burden in humans. Lead may come from water flowing through lead pipes in old dwellings or copper pipes welded with lead. Food cans are not welded with lead

Table 28. Lead content in food products purchased in Santiago metropolitan area, 1995.

Lettuce	0.3–56 μg/g wet wt (average, 12 μg/g)
Carrots	4 μg/g dry wt
Beans	4 μg/g dry wt
Potatoes	1.6 μg/g dry wt
Wheat products	< 1 μg/g dry wt
Canned evaporated milk	202 mg/L
Fruits juice, packed	> 100 mg/L

Source: Díaz and García (2003).

welding anymore, but before 2000 most canned food was packed in lead-welded cans. Due to population ignorance, once opened, food sometimes was kept in these cans, leading to increasing concentrations of lead in food. The frequent marketing of bruised or accidentally deformed cans, in which the welded layer was affected, constituted an important source of lead in food. In packaged food, lead can also originate from the packing material, especially for various beverages such as milk or fruit juice.

Few studies were performed in Chile on lead content in food. Table 28 shows data on lead in several food products, investigated at the University of Chile in 1995, and Table 29 shows lead levels in fish products measured at the University of Santiago (Díaz and García 2003).

A study directed by one of us (A.N.T.) confirmed high lead levels in lettuce cultivated near a high-traffic highway (50 m) in the coastal zone of Central Chile (Briones and Vera 2002). Samples of lettuce were taken in that site 1 yr after the ban of leaded gasoline. These displayed a high concentration of lead (50–55 μg/g), suggesting that the pollutant from leaded gasoline accumulated in the soil and was absorbed by the plants through their roots.

Recently, a study measured lead content in breakfasts and lunches given at a primary school in the VII Region (300 km south of Santiago) within a governmental program of food assistance to economically deprived children (Bastías et al. 2004). Each breakfast ration contained 1.3 μg/g lead and each lunch portion contained 0.6 μg/g lead, which totaled 714.1 μg lead/d. This contamination clearly exceeded the recommendation of the Mixed Expert Committee

Table 29. Lead content in fish species or food products purchased in Santiago metropolitan area.

Merluccius gayi	0.72–0.94 μg/g wet wt
Truchurus murpyi	1.16–1.41 μg/g wet wt
Fish flour	0.46 μg/g dry wt
Fish oil	0.13 μg/g dry wt
Canned fish	0.57 μg/g dry wt

Source: Díaz and García (2003).

from the FAO/WHO experts on food additives (Comité Mixto FAO/OMS de Expertos en Aditivos Alimentarios 1999) of a maximum of 25 μg lead/kg body wt/wk.

In Chile, regulations on lead in food (D 977/96 Ministry of Health, Chile 1996; D 475/99 Ministry of Health, Chile 1999; D 238/2000 Ministry of Health, Chile 2000) are not precise and are also insufficient, revealing a disparity in legislative criteria (Bastías et al. 2004). There is no regulation forbidding the use of lead in food can welding; although it is assumed is no longer done, there is no regulation to that effect. According to food sanitary regulations (D 977/96 Ministry of Health, Chile 1996; D 475/99 Ministry of Health, Chile 1999; D 238/2000 Ministry of Health, Chile 2000), metals in contact with food should contain less than 1% of the sum of the following elements (among others); lead, copper, chromium, antimony, tin, and/or iron. Metal cans with canned food contain higher amounts of at least one of these elements (Fe), and there are no controls to check whether dangerous elements are in higher amounts than those allowed by regulation.

VIII. Lead in Soil

Few studies were done to analyze lead content in soils of different Chilean regions. Most of these were done locally to ascertain soil levels within the areas influenced by different sources of pollution, mainly before actions intended to clean up these sites, or in places such as Arica or Antofagasta, where the effects of lead contamination on humans is evident.

A study investigated lead concentration in soil in the vicinity of several rivers. In some, lead was considered to originate from natural sources, while in others it is anthropogenic. Figure 19 shows lead content in these soils (CONAMA Chile 1994; González 1994). The profile shows that variation of values for each

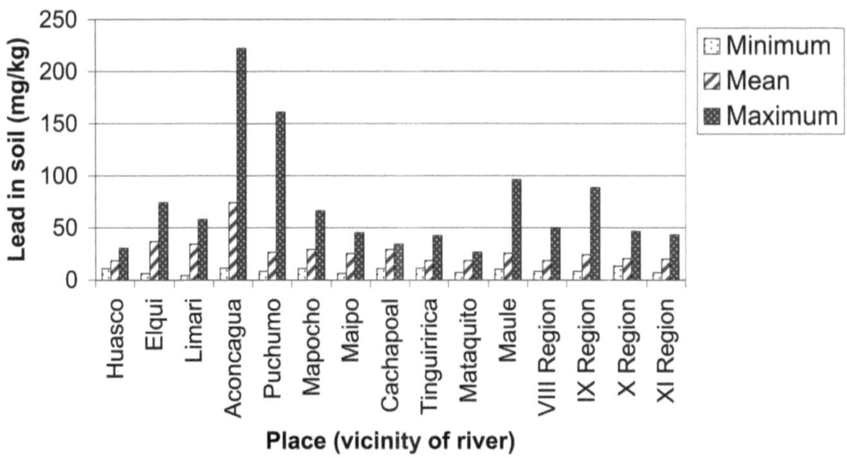

Fig. 19. Lead content in soils in vicinity of various Chilean rivers. (From CONAMA Chile 1994; González 1994.)

Table 30. Lead concentration in soils located near[a] or far[b] from industrial pollution sources (copper smelters) in the V Region (north and northwest of Santiago).

Agroecological unit	Sector	Cu (mg/kg)	Pb (mg/kg)	Zn (mg/kg)
Aconcagua	Catemu[a]	319	29	59
Aconcagua	Nogales[b,c]	135	8	28
Aconcagua	The rest[b]	63	8	89
Puchuncavi	Less than 3 km from smelter[a]	281	115	144
Puchuncavi	More than 3 km from smelter[b]	38	40	110

[c]Nogales is located far from polluting source but was contaminated by a collapse of a reservoir containing mineral wastes.
Source: González (1994).

ecosystem is low. Within the agroecological unit of the Aconcagua Valley (100 km north of Santiago), it was found that variation of lead levels in soil, especially the mean levels and the absolute maximal values of lead, coincided with those of copper (results not shown), suggesting anthropogenic contamination with both metals. In the sectors of Puchuncavi (Puchumo) and Nos (Maipo valley), high lead concentrations were from anthropogenic sources, associated with industrial processes. Puchuncavi is in the area of copper smelter emission of large amounts of sulfur dioxide and arsenic, together with other heavy metals. A decrease in concentration of lead at lower valley locations may result from dilution of the valley river water by cleaner river affluent.

Table 30 shows mean values of various heavy metals in soils (0–20 cm) at sites located near and far from contamination sources. Copper smelters are located in Catemu and Puchuncavi. Nogales suffered from a collapse of a reservoir dam containing mineral wastes. In some places, lead content in soil increases, presumably due to combination with other molecules present in soil and its precipitation as nonsoluble compounds.

These studies concluded that in many places the total lead content results from natural origin but in others it may be anthropogenic. There was no relation, however, between increased lead content in soil and the existence of lead mineral zones such as the Maipo Valley and the XI Region, suggesting that these minerals do not naturally dissolve in water.

IX. Recommendations

Exposure to lead is known to affect various organs and systems in humans and in experimental animals. In adults, exposure to moderate lead levels can cause damage that can be reversed, at least in part, after decreasing lead levels in the blood and an additional recovery time. Exposure of children, and especially during the neonatal period of life and during fetal stages of intrauterine development, even at much lower lead concentrations, causes irreversible changes that persist through life. This is the basis for recommending extreme caution to avoid lead exposure of children, of babies during the early stages of their development, and of fetuses during the last months of intrauterine life.

This is also the reason why norms and regulations for labor lead exposure levels or for maximal allowed lead concentrations in blood should be much lower for females than for males. Taking into consideration that the organism does not distinguish between lead and calcium, in a lead-exposed population lead will accumulate in bones together with calcium. In contrast to males, females become pregnant and obtain calcium for fetal needs from their own bone reserves, so that if lead content in her bones is elevated, the mother will withdraw calcium together with lead, which will affect the fetal tissues. Further, during breastfeeding, mothers draw additional calcium needed for the child's growth, and if lead is accumulated in maternal bones, it will result in newborn exposure to this toxic element.

Based on these considerations, we recommend: (a) further changes in lead regulations, (b) formal educational programs, (c) informal educational campaigns through mass communication media, (d) high-level educational campaigns for the political sector, public administration, and the enterprise management sector, (e) administrative actions to decontaminate polluted areas, and (f) medical actions aimed mainly at detection and therapeutics, with special consideration of nutritional intervention and special care for pregnancy, early postnatal life, and childhood. In addition, we recommend international agreements regulating commerce of products dangerous to health, including lead.

New Chilean regulations on lead content in air and paints are described above. Both are significant advances and are similar to those recommended by the World Health Organization (WHO) (lead in air) or in force in Europe (lead in air) or the U.S. (lead in paints). The limits of biological tolerance to lead were established in Chile in 1992 by Decree 745/92, art. 101, as 50 μg/dL lead in blood (D 745/92, Ministry of Health, Chile 1992). In 1999, stricter limits of biological tolerance were established by Decree 594, art. 113 (D 594/99, Ministry of Health, Chile 1999), decreasing tolerance for males to 40 μg/dL and for females to 30 μg/dL. These changes generated polemics at the political level and pressure from economically affected sectors to lessen regulations, and a new regulation established by Decree 201/2001, art. 17 (D 201/2001, Ministry of Health Chile, 2001) again changed the limits of biological tolerance, increasing it for females to 40 μg/dL and keeping the 40 μg/dL value for males. The Colegio Médico de Chile submitted a critical analysis with all scientific arguments to reverse that decree (Tchernitchin and Castro 2001), but no satisfactory results were obtained. We strongly recommend improving regulations on biological tolerance aimed fundamentally at the protection of children and women. We further recommend improving norms on lead content in drinking water and drawing regulations on lead content in food and in several products accessible to children. All these measures will be useless if these regulations are not accompanied with improved control measures and strict penalties for transgressions, both of which are deficient in Chile.

Formal education programs on health and environmental effects of lead and other pollutants are important to change human behavior for prevention and mitigation. Recently programmed changes in education by the government mainly aimed at environmental protection but were not sufficiently strong on environmental toxicology and health effects. In formal educational campaigns have been done since 1996 by the Colegio Médico de Chile and were sufficiently effective in changing people's attitude toward several pollutants, including lead. Recently, several nongovernmental organizations with participation of scientists or physicians also contributed to this purpose.

During recent years, the Chilean Colegio Médico implemented a course on environmental changes and health effects for television, radio, and press reporters who specialized on health and the environment. This course contributed to improving the quality and credibility of the articles they present in the media, helping to increase acceptance of the described environmental problems by citizens and also by politicians. The Colegio Médico de Chile sent technical information to several political parties and presented several lectures to transmit this knowledge to this sector. The Chilean Parliament, both the Lower Chamber (Chamber of Deputies) and the Higher Chamber (Chamber of Senators), usually requires information about lead or other pollutants from the Colegio Médico, and several presentations *in extenso* were also prepared for them. Our participation in the Commission of the Chamber of Deputies to investigate the effects of lead in Arica and propose a decontamination plan was a result of this campaign. This is a good example showing that scientists need to work together with politicians, legislators, and the educational sector to obtain results in lead decontamination as well as in other areas of environmental toxicology.

Many administrative actions were undertaken by local or central Chilean authorities under the pressure of population awareness of the damage caused by lead, following the campaign designed by the Chilean Colegio Médico. Cleaning up of contaminated areas in Arica and Antofagasta, relocation of toxic wastes to a desert area near Arica, plans to transfer the port from which the lead mineral concentrates are shipped from Antofagasta to Mejillones and to declare the zone of Antofagasta as a National Catastrophe Zone to obtain funds to make these changes, all were these actions taken under the social and political pressure of the population who for the first time felt damaged by lead exposure. From this it can be concluded that changes favoring health of the population are done most efficiently and rapidly if there is pressure from the population, and that arguments are difficult to refute when based on scientific knowledge. Therefore, the best way to solve environmental or health problems is to communicate them to the entire population, providing valid and solid scientific arguments, and await population pressure on the responsible authorities.

Medical actions need to focus mainly on the detection of lead exposure and to therapeutic procedures to treat or to prevent sequelae. It is necessary to take special consideration for those groups that are more vulnerable to lead effects or develop irreversible changes which persist through life: newborns or babies during their early postnatal life, children, and pregnant mothers, to avoid damage of their fetuses. Therefore, it is necessary to develop methods for screening and detecting affected subjects with inexpensive methods, to avoid further exposure, to decrease blood lead levels, or to treat the effects. Prevention methods may involve changes in diet, especially calcium and iron supplementation and decrease in lipid ingestion.

It is necessary to take into consideration the most conspicuous effects of lead exposure: decrease in intelligence coefficient and in attention and memory capacity, hyperactive behavior, aggressive behavior, and delinquent behavior in adolescence. Therefore, all schoolchildren with these characteristics in the contaminated areas (Arica, Antofagasta) should be investigated for blood lead content. It is highly recommended that mothers planning to get pregnant have their blood lead levels screened so that they may be treated with pharmaceutical therapy (including chelation therapy) before the initiation of pregnancy. All pregnant women should measure their blood lead levels to decide a therapy with less risk for the child such as calcium or vitamin supplementation, besides the need to identify the lead source to avoid further exposure. After delivery, blood should be analyzed for lead level measurement in both mothers and newborns to decide on the convenience of breastfeeding or artificial feeding. All children from families with history of lead exposure should be screened periodically to prevent irreversible damage by lead.

The hyperactive or aggressive behavior induced by early exposure to lead may be irreversible. This behavior risks the development of antisocial or delinquent behavior later in adolescence. We propose that this aggressive behavior can be sublimated, under psychological treatment, to other aggression-like activities, such as sports, dance, hobbies, or many other activities. Therefore, we recommend hiring psychologists at schools so that they can help children to sublimate this aggressive behavior and to channel it to socially positive behavior and avoid children transforming into a delinquent population.

Finally, we recommend the implementation of international agreements regulating commerce of products containing components dangerous for health. Governments of different countries should agree to prohibit export of such products, those containing lead for instance,when their constituents are prohibited or regulated in the countries of their origin, because of ethical reasons even when there is no such regulation in the importing country.

Summary

Lead is a very toxic environmental pollutant that affects people living or working in contaminated areas or ingesting this element. Since the beginning of lead use, it has left evidence of lead poisoning with dramatic effects on the destiny of ancient civilizations. It was recently proposed that this toxicity can also contribute to the decline of current societies through lead-induced impairment of intelligence, increased tendency to abuse of drugs, psychological and behavioral changes, and increase in delinquent behavior.

The effects of chronic exposure to lead in adolescents or adults may be reversed, at least in part, after a decrease in the blood lead levels. In contrast, in children, especially at earlier ages, the effects are irreversible, may persist through life, and some of them may be induced by the mechanism of imprinting. Exposure to lead during the perinatal period of life may occur in mothers previously exposed to the metal, presenting increased concentration of lead in bones; fetuses and newborns are most vulnerable to the toxic effects of lead because the damage persists throughout life.

The various sources of lead exposure in Chile as well as the evolution of regulations and mitigation measures are reviewed. An important source of contamination with lead was the existence of paints with high lead content, used to paint house walls, children's furniture, and toys, among other objects. The highest content of lead in paints was not in those bearing Chilean trademark names, but from a foreign trademark with formulas licensed to produce in Chile, although the same products in the country of origin (U.S.) did not present measurable lead levels. From 1997 on, new legislation regulated lead content in paints. Many houses, furniture, children's toys, and other objects painted before 1997 are still an important source of exposure to lead, mainly for children.

Formerly, the widest source of lead contamination in Chile was the use of leaded gasoline, mainly in greatly populated cities. There has been a progressive decrease in lead content in leaded gasoline, which was banned in April 2001. Lead pollution still persists in highly populated cities as city soil, home soil, and as soil contamination near highways.

Clusters of different magnitude originated from point sources, of which the most relevant in magnitude was that which occurred in Ñuble with the use of wheat flour contaminated with lead, caused by the use of a mill whose stone was repaired by lead welding. Several other clusters of smaller magnitude are frequently caused by the widespread practice of battery repair and recovery by small enterprises or as a familial productive activity, affecting all the family group and occasionally persons living in the vicinity.

Two special cases of very important environmental contamination occurred, affecting the population of two major cities. The most dramatic case was the storage of powdered lead mineral concentrates at the ports or the railroad terminals within the cities of Arica and Antofagasta, where it usually remained until shipped to other countries. Lead from this source is the most relevant source of exposure in the cities of Arica and Antofagasta. Another source of lead, which affected the population of Arica, originated from toxic wastes imported by Promel from the Swedish company Boliden Metal, stored without protection in Arica suburbs, where a few years later new dwellings were constructed and residents were exposed for more than 10 years.

Occupational exposure frequently occurs in painters, welders, and mining or smelting workers. Various degrees of food lead contamination may increase exposure to the metal. Vegetables grown in or near highly populated areas, lead pipes in old dwellings or copper pipes welded with lead, and formerly lead from canned food, especially those that had been accidentally deformed or were opened and the food stored in those cans.

Little information exists on lead content in soil. There are sites where lead concentrations are increased from anthropogenic sources. In regions known to have abundant lead minerals, soil is not contaminated in the vicinity of rivers, suggesting that raw lead minerals do not dissolve in water and do not affect the soil.

Acknowledgments

Financed in part by funds from the Colegio Médico de Chile; the Arica study was financed in part by Fund of the Americas grant. Lead measurements in blood and in household paints by the Institute of Public Health ISP from the Chilean Ministry of Health are acknowledged. Technical help from Gladys Cisternas, Bernardita Araya, Dora Méndez, and Leonardo Figueroa is appreciated.

References

Al-Hakkak ZS, Hamamy HA, Murad AMB, Hussain AF (1986) Chromosome aberrations in workers at a storage battery plant in Iraq. Mutat Res 171:53–60.

Alonso P, Castro H, Davis ME, Goza J, Hargous J, Rivera L, Tassara E (1997) Intoxicación por plomo: estudio epidemiológico Ñuble, Chile 1996. EPI Visión, Ministerio de Salud, Chile, pp 10–15.

Banks EC, Ferretti LE, Shucard DW (1997) Effects of low level lead exposure on cognitive function in children: a review of behavioral, neuropsychological and biological evidence. Neurotoxicology 18:237–282.

Baker EL, Landrigan PJ, Barbour AG, Cox DH, Folland DS, Ligo RN, Throckmorton J (1979) Occupational lead poisoning in the United States: clinical and biochemical findings related to blood lead levels. Br J Ind Med 36:314–322.

Bastías JM, Muñoz O, Moreno I, Rubilar C (2004) Evaluación del contenido de arsénico y plomo en raciones de desayuno y almuerzos entregados a escolares de la VII región, Chile. CD Abstracts, XIII Latin American and Caribean Seminar on Food Science and Technology, Montevideo, Uruguay.

Bornschein RL, Grote J, Mitchell T, Succop PA, Dietrich KN, Krafft KM, Hammond PB (1989) Effects of prenatal lead exposure on infant size at birth. In: Smith MA, Grant LD, Sors AI (eds) Lead Exposure and Child Development. An International Assessment. Kluwer, Dordrecht, pp 307–319.

Briones MJ, Vera PA (2002) Prospección y caracterización de los metales pesados en Chile, un caso particular: plomo en hortalizas cultivadas en la Provincia de San Antonio, V Región. Thesis, UTEM, Santiago, Chile.

Byers RK, Lord E (1943) Late effects of lead poisoning on mental development. Am J Dis Child 66:471–494.

Cardenas A, Roels H, Bernard AM, Barbon R, Buchet JP, Lauwerys RR, Rosello J, Ramis I, Mutti A, Franchini I, Fels LM, Stolte H, de Broe ME, Nuyts GD, Taylor SA, Price RG (1993) Markers of early renal changes induced by industrial pollutants. II. Application to workers exposed to lead. Br J Ind Med 50:28–36.

Cisternas R, Sáez M (1980) Liberación de delta-ala en orina de niños: un indicador de contaminación por plomo en Santiago. Rev Chil Pediatr 51:13–18.

Cohen N, Modai D, Golik A, Weissgarten J, Peller S, Katz A, Averbukh Z, Shaked U (1989) Increased concanavalin A-induced suppressor cell activity in humans with occupational lead exposure. Environ Res 48:1–6.

Comité Mixto FAO/OMS de Expertos en Aditivos Alimentarios (1999) Evaluación de ciertos aditivos alimentarios y contaminantes de alimentos. 53° Informe de JECFA. N°896, Ginebra.

CONAMA Chile (1994) Perfil Ambiental de Chile. Comisión Nacional de Medio Ambiente (CONAMA), Chile.

Cooper WC (1988) Deaths from chronic renal disease in US battery and lead production workers. Environ Health Perspect 78:61–63.

Csaba G (1980) Phylogeny and ontogeny of hormone receptors: the selection theory of receptor formation and hormonal imprinting. Biol Rev 55:47–63.

Csaba G, Inczefi-Gonda A, Dobozy O (1986) Hormonal imprinting by steroids: a single neonatal treatment with diethylstilbestrol or allylestrenol gives a rise to a lasting decrease in the number of rat uterine receptors. Acta Physiol Hung 67:207–212.

Davis JM, Grant LD (1992) The sensitivity of children to lead. In: Guzelian PS, Henry CJ, Olin S (eds) Similarities and Differences Between Children and Adults: Implications for Risk Assessment. ILSI Press, Washington, DC, pp 15–27.

Davis JM, Svendsgaard DJ (1990) Nerve conduction velocity and led: a critical review and meta-analysis. In: Johnson BL, Anger WK, Durao A, Xinteras C (eds) Advances in Neurobehavioural Toxicology: Applications in Environmental and Occupational Health. Lewis, Chelsea, MI, pp 353–376.

Deknudt G, Leonard A, Ivanov B (1973) Chromosome aberrations observed in male workers occupationally exposed to lead. Environ Physiol Biochem 3:132–138.

Díaz O, García M (2003) Avances en Toxicología de Contaminantes Químicos en Alimentos. Programa Iberoamericano de Ciencia y Tecnología para el Desarrollo, Santiago, Chile.

Dietrich KN, Succop PA, Berger OG, Keith RW (1992) Lead exposure and the central auditory processing abilities and cognitive development of urban children: the Cincinnati lead study cohort at age 5 years. Neurotoxicol Teratol 14:51–56.

Dobozy O, Csaba G, Hetényi G, Shahin M (1985) Investigation of gonadotropin thyrotropin overlapping and hormonal imprinting in the rat testis. Acta Physiol Hung 66:169–175.

D 374/97 Ministry of Health, Chile (1997) Establece norma sobre contenido de plomo en pinturas. Ministry of Health, Republic of Chile.

D 594/99 Ministry of Health, Chile (1999) Reglamento sobre condiciones sanitarias y ambientales básicas en los lugares de trabajo. Ministry of Health, Republic of Chile, art 113.

D 201/2001, Ministry of Health, Chile (2001) Modifica reglamento sobre condiciones sanitarias y ambientales básicas en los lugares de trabajo. Ministry of Health, Republic of Chile, art 17.

D 745/92 Ministry of Health, Chile (1992) Reglamento sobre condiciones sanitarias y ambientales básicas en los lugares de trabajo. Ministry of Health, Republic of Chile, art 101.

D 977/96 Ministry of Health, Chile (1996) Reglamento sanitario de los alimentos. Ministry of Health, Republic of Chile.

D 475/99 Ministry of Health, Chile (1999) Modificación del reglamento sanitario de los alimentos. Ministry of Health, Republic of Chile.

D 238/2000 Ministry of Health, Chile (2000) Modificación del reglamento sanitario de los alimentos. Ministry of Health, Republic of Chile.

DS 136/00 Chile (2000) Establece norma de calidad primaria para plomo en el aire. Ministry of the General Secretariat of the Presidency, Republic of Chile.

DS 058/03 Chile (2003) Reformula y actualiza plan de prevención y descontaminación atmosférica para la Región Metropolitana, Ministry of the General Secretariat of the Presidency, Republic of Chile.

Ewers U, Stiller-Winkler R, Idel H (1982) Serum immunoglobulin, complement C3, and salivary IgA level in lead workers. Environ Res 29:351–357.
Frenz P, Vega J, Marchetti N, Torres J, Kopplin E, Delgado I, Vega F (1997) Exposición crónica a plomo ambiental en lactantes chilenos. Rev Méd Chile 125:1137–1144.
Gallardo L, Olivares G, Aguayo A, Langner J, Aarhus B, Engardt M, Gidhagen L (2000) Regional dispersion of oxidized sulfur over Central Chile using the HIRLAM-MATCH system. Comision Nacional de Medio Ambiente (CONAMA), Chile.
Gilfillan SC (1965) Lead poisoning and the fall of Rome. J Occup Med 7:53–60.
González S (1994) Geoquímica de metales pesados en Chile. In: Impacto Ambiental de Metales Pesados en Chile. Simposio sobre Contaminación Ambiental, Santiago, Chile, pp 10–29.
Grandjean P, Jensen BM, Sando SH, Jorgensen PJ, Antonsen S (1989) Delayed blood regeneration in lead exposure: an effect on reserve capacity. Am J Public Health 79:1385–1388.
Graziano JH, Slavkovic V, Factor-Litvak P, Popovac D, Ahmedi X, Mehmeti A (1991) Depressed serum erythropoietin in pregnant women with elevated blood lead. Arch Environ Health 46:347–350.
Jaremin B (1983) Blast lymphocyte transformation (LTT), rosette (E-RFC) and leukocyte migration inhibition (MIF) tests in persons exposed to the action of lead during work. Report II Bull Inst Marit Trop Med (Gdynia) 34:187–197.
Koller LD (1990) The immunotoxic effects of lead in lead-exposed laboratory animals. Ann NY Acad Sci 587:160–167.
Lang DS, Meier KL, Luster MI (1993) Comparative effects of immunotoxic chemicals on in vitro proliferative responses of human and rodent lymphocytes. Fundam Appl Toxicol 21:535–545.
Long GJ, Rosen JF (1994) Lead perturbs 1,25-dihydroxyvitamin D_3 modulation of intracellular calcium metabolism in clonal rat osteoblastic (ROS 17/2.8) cells. Life Sci 54:1395–1402.
Mahaffey KR, Rosen JP, Chesney RW, Peeler JT, Smith CM, DeLuca HF (1982) Association between age, blood lead concentration, and serum 1,25-dihydrocalciferol levels in children. Am J Clin Nutr 35:1327–1331.
Ministry of Health, Chile (1993) Blood lead means and percent $\geq 10\,\mu$g/dl maternal, umbilical cord, 6 m, 12 m and 18 m, Santiago and San Felipe 1992 cohorts. Ministry of Health, Chile.
NAS-NRC (1972) Airborne Lead in Perspective. National Academy of Science, Washington, DC.
Needleman H, Gunnoe C, Leviton A, Reed M, Peresie H, Maher C, Barrett P (1979) Deficits in psychological and classroom performance of children with elevated dentine lead levels. N Engl J Med 300:689–695.
Needleman HL, Rabinowithz M, Leviton A, Linn S, Schoenbaum S (1984) The relationship between prenatal exposure and congenital anomalies. JAMA 251:2956–2959.
Needleman HL, Riess JA, Tobin MJ, Biesecker GE, Greenhouse JB (1996) Bone lead levels and delinquent behavior. JAMA 275:363–369.
Ong CN, Endo G, Chia KS (1987) Evaluation of renal function in workers with low blood lead levels. In: Occupational and Environmental Chemical Hazards: Cellular and Biochemical Indices for Monitoring Toxicity. Horwood, Chicester, pp 327–333.
Pagliuca A, Mufti GJ, Baldwin D, Lestas AN, Wallis RM, Bellingham AJ (1990) Lead poisoning: clinical, biochemical, and haematological aspects of a recent outbreak. J Clin Pathol 43:277–281.

Regional Public Health Service of Antofagasta (2000) Informe evaluación Playa Carboncillo, Antofagasta, Chile.

Ronis MJJ, Badger TM, Shema SJ, Roberson PK, Shaikh F (1996) Reproductive toxicity and growth effects in rats exposed to lead at different periods during development. Toxicol Appl Pharmacol 136:361–371.

Ruthllant J, Garreaud R (1995) Meteorological air pollution potential for Santiago, Chile: towards an objective episode forecasting. Environ Monit Assess 34:223–244.

Schmid E, Bauchinger M, Pietruk S, Hall G (1972) Cytogenic action of lead in human peripheral lymphocytes in vitro and in vivo (in German). Mutat Res 16:401–406.

Schwartz J, Otto D (1991) Lead and minor hearing impairment. Arch Environ Health 46:300–305.

Schwartz J, Angle C, Pitcher H (1986) Relationship between childhood blood lead levels and stature. Pediatrics 77:281–288.

Schwartz J, Landrigan PL, Baker EL (1990) Lead-induced anemia: dose-response relationships and evidence for a threshold. Am J Public Health 80:165–168.

Sepúlveda V, Vega J, Delgado I (2000) Exposición severa a plomo ambiental en una población infantil de Antofagasta, Chile. Rev Méd Chile 128:221–232.

Servicio de Salud de Arica (2000) Resultados exámenes en Población Villa Santa María. Plomo en Sangre. Ministerio de Salud, Chile, Servicio de Salud Arica, Subdirección Médica, Unidad de Epidemiología.

Servicio de Salud de Arica (2001) Nómina de personas con muestras de plomo en sangre con niveles $>10\,\mu$g/dl confirmada por I.S.P. Muestreo realizado en año 2000 y controles en 2001. Población industriales II, III y IV. Pob. Co. Chuño. Ministerio de Salud, Chile, Servicio de Salud Arica, Subdirección Médica, Unidad de Epidemiología.

Servicio Municipal de Salud Arica (2001) Informe Julio 2001. Campaña de Salud Niños de Arica Libres de Plomo. Municipalidad de Arica, República de Chile.

SESMA Chile (2001) Determinación de niveles de exposición a plomo en el aire de la Región Metropolitana entre 1997 y 2000, Servicio de Salud del Ambiente de la Región Metropolitana SESMA, Gobierno de Chile.

SESMA Chile (2002) Caracterización de elementos inorgánicos presentes en el aire de la Región Metropolitana 1997–2000, Servicio de Salud del Ambiente de la Región Metropolitana SESMA, Gobierno de Chile.

Silva H, Tchernitchin AN, Aguilera L (1999) Tiempo de reacción ante estímulos visuales y su relación con los accidentes del tránsito. In: IV Congreso Latinoamericano de Epidemiología, Santiago, Chile, p 131.

Staessen JA, Christopher JB, Fagard R, Lauwerys RR, Roels H, Thijs L, Amery A (1994) Hypertension caused by low-level lead exposure: myth or fact? J Cardiovasc Risk 1: 87–97.

Tchernitchin AN (2001) Efectos diferidos de la exposición prenatal, neonatal o durante el desarrollo infantil a contaminantes ambientales. Visión Médica Regional (Concepción, Chile) 6 (5):76–83.

Tchernitchin AN, Castro JL (2001) Oficio de Presidencia Colegio Médico, N°0994, a Ministra de Salud.

Tchernitchin AN, Castro JL (2002) Carta al Editor, Respuesta al Gerente Técnico Sherwin Williams Chile SA. Visión Médica Regional (Concepción, Chile) 6 (7):93–97.

Tchernitchin AN, Tchernitchin N (1992) Imprinting of paths of heterodifferentiation by prenatal or neonatal exposure to hormones, pharmaceuticals, pollutants and other agents or conditions. Med Sci Res 20:391–397.

Tchernitchin AN, Villarroel C (2002) Colegio Médico (Visiones sectoriales del CDS). CDS Chile 1:42–43.

Tchernitchin AN, Villarroel C (2003) El derecho a la salud. In: Godoy Y, Carrasco D (eds) Derechos Económicos, Sociales y Culturales en Chile. Informe de la Sociedad Civil. Servimpress, Santiago, Chile, pp 45–61.

Tchernitchin AN, Villagra R, Tchernitchin NN (1997) Effect of chronic exposure to lead on immature rat leukocytes. Med Sci Res 25:355–357.

Tchernitchin NN, Tchernitchin AN, Mena MA, Villarroel C, Guzmán C, Poloni P (1998) Effect of subacute exposure to lead on responses to estrogen in the immature rat uterus. Bull Environ Contam Toxicol 60:759–765.

Tchernitchin AN, Tchernitchin NN, Mena MA, Unda C, Soto J (1999) Imprinting: perinatal exposures cause the development of diseases during the adult age. Acta Biol Hung 50:425–440.

Tchernitchin NN, Clavero A, Mena MA, Unda C, Villagra R, Cumsille M, Tchernitchin AN (2003) Effect of chronic exposure to lead on estrogen action in the prepubertal rat uterus. Environ Toxicol 18:268–277.

Tuppurainen M, Wagar G, Kurppa K, Sakari W, Wambugu A, Froseth B, Alho J, Nykyri E (1988) Thyroid function as assessed by routine laboratory tests of workers with long-term lead exposure. Scand J Work Environ Health 14:175–180.

USEPA (1986) Air quality criteria for lead. EPA-600/8-83/028aF-dF. U.S. Environmental Protection Agency, Washington, DC.

USEPA (1990) Supplement to the 1986 EPA air quality criteria for lead. Volume 1, Addendum: Relationship of blood pressure to lead exposure. EPA/600/8-89/0 49F. U.S. Environmental Protection Agency, Office of Health and Environmental Assessment, Washington, DC.

Villagra R, Tchernitchin NN, Tchernitchin AN (1997) Effect of subacute exposure to lead and estrogen on immature pre-weaning rat leukocytes. Bull Environ Contam Toxicol 58:190–197.

Waldron HA (1973) Lead poisoning in the ancient world. Med Hist 17:391–399.

Ward NI, Watson R, Bryce-Smith D (1987) Placental element levels in relation to fetal development for obstetrically normal births. A study of 37 elements. Evidence for the effects of cadmium, lead, and zinc on fetal growth and for smoking as a source of cadmium. Int J Biosoc Res 9:63–81.

Weeden RP, Maesaka JK, Weiner B, Lipat GA, Lyons MM, Vitale LF, Joselow MM (1975) Occupational lead nephropathy. Am J Med 59:630–641.

Weeden RP, Mallik DK, Batuman V (1979) Detection and treatment of occupational lead nephropathy. Arch Intern Med 139:53–57.

Winder C (1993) Lead, reproduction and development. Neurotoxicology 14:303–317.

Manuscript received January 17; accepted January 23, 2005.

Rev Environ Contam Toxicol 185:141–177

Human Nails as a Biomarker of Element Exposure

A. Sukumar

Contents

Contributed by George W. Ware.

A. Sukumar (✉)

Department of Education in Science and Mathematics, Regional Institute of Education, National Council of Educational Research and Training, Mysore-570 006, India.

I. Introduction

The concept of biological monitoring has been practiced by the medical profession for decades (Waritz 1979); however, the formal application of this technique on a broad scale to industrial hygiene is attributed to Elkins (1954). Elkins subsequently attempted to correlate exposure level and urinary level with toxicity of chemicals. Since then, the concept of biological monitoring for trace elements and xenobiotics has been extended to include monitoring of the bile and feces, exhaled air, blood, perspiration, tears, nails of fingers and toes, hair, milk, teeth, and saliva. The order of preference for current use of such samples is blood, urine, hair, nails, teeth, milk, saliva, tears, perspiration, exhaled air, biliary, and fecal substances, skin scales, and other.

In comparison to other biological samples, there are many advantages in utilizing nails as a biomarker of toxic element exposure and nutritional mineral status (Chen et al. 1999). These include (1) the ease with which nails can be sampled, stored, transported, and handled (Daniel et al. 2004); (2) standardized methods are available for collection, washing, preparation, and sophisticated instrumental analysis with use of quality control samples for precision and accuracy (Chaudhary et al. 1995; Rayman et al. 2003; Chandra Sekhar et al. 2003; Platz et al. 2002; Bergomi et al. 2002); (3) storage of aliquots of sample is simple for reanalysis and toenail measure of arsenic (As) is reproducible over a period of several years (Garland et al. 1994); (4) once elements are incorporated into keratin of nails, the levels remain isolated from other metabolic activities in the body with no fluctuation in element levels due to changing body metabolic activities, unlike the blood; (5) nails, which conserve the pattern of trace element composition beyond the life span of the cells and because of their resistance to decay, are exceptionally well suited for long-distance transport; also, a large number of specimens can be procured and stored, a requirement for population studies; (6) toenails take several months to a year to grow out and are used for the measure of past exposure (Longnecker et al. 1993; Karagas et al. 2000); (7) a number of trace elements are considerably more concentrated in nails than in urine or blood (Karpas 2001); and (8) the ability to analyze nail content of several elements simultaneously is a definite advantage, especially in view of the growing knowledge of trace element interaction (Rodushkin and Axelsson 2000a).

For the past three decades, nails have been employed along with hair in measuring elemental status for nutritional evaluation, disease diagnosis, and exposure profile from place of work or the environment (Mahler et al. 1970) because they are similar in structural composition and development (Vance et al. 1988). They are the metabolic end product of skin and contain semihard, cornified cells with trace elements in a similar proportion as in the cells from which they are produced (Hopps 1977). Thus, alteration of element composition of cells is reflected by the pattern of element distribution of the nails and hair.

Literature surveys reveal that nails are not analyzed for element content as extensively as hair samples (Vance et al. 1988). Even then, nails are preferable to hair for the following reasons. (1) Generally, nails are less susceptible to external contamination than hair (Karagas et al. 2000), as the levels of elements vary along the length of hair, and area of exposure to external contamination is greater in hair than in nails; (2) selenium levels in toenails (Se-TN) are highly reproducible in two sets of specimen collected about a year apart (Krogh et al. 2003); (3) nail clipping is relatively simpler than cutting hair at the base of scalp, and thus a high participation rate of 97% in a study has been reported by Karagas et al. (2000); (4) the number of biological factors (length, color, types, growth rate, and cycle) of hair influencing elemental levels (Sukumar 2002) is numerous, whereas the biological factors of nails are at a minimum; (5) the standard method of collection, transport, storage, washing, digestion, and instrumental analysis is more or less similar for both kinds of samples; (6) nails become a useful substitute biomarker of element exposure, as untreated control hair is not available; and (7) interpersonal variability in As levels is lower in nails than in hair (Hinwood et al. 2003).

In view of these factors, analysis of human nail samples for element content is an increasing trend in diverse research fields (Takagi et al. 1988). Such analyses, in spite of having characteristics unique to a particular region, subject group, exposure status, and specific purpose, are compared among reported studies at international levels to derive a correlation between element status and health in the general population, workers, or patients. To validate analysis of nails as an indicator of body burden of minerals, earlier and recent data on human nails are reviewed, focused on analytical methods and factors influencing nail levels of elements. Moreover, the use of nail analysis for biomonitoring of environmental and occupational exposures and health status is discussed with a critical evaluation of its use. Emphasis is given to examining nail analysis with the perspective of significance of element interaction, speciation, supplementation in nails, and the correlation of element levels of nails with other samples.

II. Analysis

An indispensable purpose for nail analysis is that elements measured are indicative of accumulation during nail formation and thus depict the internal body store of elements (Cheng et al. 1995; Gibson 1989). After formation, nails are regarded as dead material and are also exposed to outside sources of elements for longer periods. Hence, analysis of elements must include procedures that eliminate the ambiguity of external contamination and reflect exclusively internal element uptake.

The most important procedures generally adopted during analysis are (1) sampling with questionnaire, (2) washing, (3) desiccation, (4) digestion, (5) instrumental estimation of element concentrations, (6) uses of quality control samples, and (7) interpretation with statistical analysis.

A. Sampling

Compared to hair, sampling of nails is easier and biological factors of nails influencing levels of elements are fewer in number. Normally, nails are sampled from either the thumb or great toe. The best qualities of nails as a biomarker are rapid growth, less external contamination, adequate sample availability, and incorporation of elements in the tissue. Small quantities of nail sample are sufficient for measurement with nuclear activation analysis (NAA); hence, sample of a single nail of either thumb or great toe is preferred to samples of all the nails. Kanabrocki et al. (1979) used thumbnails for estimating the concentrations of zinc (Zn), chromium (Cr), Se, silver (Ag), mercury (Hg), gold (Au), and cobalt (Co) using the thermal neutron activation technique. Similarly, Rogers et al. (1993) employed clippings taken from nails of both great toes to determine concentrations of iron (Fe), Zn, calcium (Ca), Cr, and Co in 661 cases and 466 controls from western Washington. If the nails are sampled from all fingers or toes, the amount available is enough for measurement with use of atomic absorption spectrophotometry (AAS) or inductively coupled plasma spectrophotometry (ICP). Therefore, the amount of sample required for element estimation determines the selection of instrumental analysis or vice versa.

In a majority of cases, nail sampling is carried out with use of stainless steel nail clippers, whereas ceramic blade cutters are employed to collect nails of subjects with the idea that the stainless steel cutter may contribute to the external source of elements (Chandra Sekhar et al. 2003). McCurdy and Hindmarsh (1987) collected nail clippings from donors using Teflon-coated stainless steel scissors. Nails that have been polished are not usually sampled because elements added in the nail polish may become an external element source.

B. Exposure Time and Sampling Time

Nails incorporate elements during formation and the length of nails clipped at time of sampling is of different size, so that the levels of elements analyzed in the samples may indicate varying periods of accumulation. Hence, Yoshizawa et al. (2002) showed that nail clippings taken from all 10 fingers or toes at a time reflected the incorporation of elements, particularly Hg, that had occurred over approximately 1 yr. Hunter et al. (1990a) corroborated that nails from all 10 toes are likely to represent exposure status occurring over a 2- to 12-mon period. Similarly, Martin-Moreno et al. (2003) found that Zn levels in toenails showed the level occurred from dietary Zn intake over 3–12 mon. Morris et al. (1983) found that analysis of nails from the large toe provided an integrated measure of iodine (I) exposure over a 2- to 4-wk period approximately 1 yr before clippings.

Time of sampling nails among various groups or between controls and corresponding cases becomes an essential factor, resulting in variation in element levels. Sampling should be carried out simultaneously or the time gap among groups must be narrow, because nail levels of elements indicate past exposure conditions that are further correlated to exposure period and ill health (Mannisto et al. 2000).

C. Washing

Nail clippings are subjected to a standard washing procedure for removal of any exogenous source of trace elements, because the ultimate goal of washing is to completely remove loosely adhering external metals associated with fat, sweat, and dirt without altering the endogenous content of elements of the samples (Caroli et al. 1994). Washing involves thorough stirring of nail samples using an ultrasonic bath with three solvents in the sequence of acetone, deionized water, and 0.5% Triton X-100 water solution (Rodushkin and Axelsson 2000a). According to Chen et al. (1999), cleansing the samples once with 1% Triton X-100 for 20 min ultrasonically before analysis is satisfactory for removing surface contamination. However, mineral acids or strong complexing agents such as ethylenediaminetetraacetic acid (EDTA) should be employed with the greatest care or avoided (Gulson 1996; Caroli et al. 1992; Rodushkin and Axelsson 2000a).

Ndiokwere (1985) and Bate and Dyer (1965) found that intrinsic sources of heavy metals in hair or nails were not essentially influenced by washing owing to their great tendency to complex with disulfide groups in the keratin protein. The results from one wash versus multiple washes confirmed that nails cleansed once, twice, or thrice did not vary substantially in As content (Chen et al. 1999).

Although minimizing extrinsic contamination is a fundamental aim of washing, an extensive procedure poses the risk of extraction of elements bound to the nail matrix. Bank et al. (1981) demonstrated that treatment with organic solvents resulted in less elemental loss than with aqueous detergents, whereas aqueous acids caused the greatest loss. Rayman et al. (2003) used a toenail sample washing including a sequence of washing with acetone (1 min), deionized distilled water (3 washes of 3 min), and acetone (1 min) ultrasonically.

As the protocols of washing fall into two approaches, such as use of either detergents or organic solvents, Harrison and Tyree (1971) compared these two procedures and found that detergent washing appears more advantageous in that it corresponds more closely to *in situ* washing and deduced that use of detergents is preferable to organic solvents. Further, washing varies depending upon the group of elements analyzed and the analytical instruments used. For ICP measurement of cadmium (Cd), lead (Pb), and copper (Cu), toenails were washed using Triton

X-100 (5% in deionized distilled water) solution for 15 min ultrasonically and then with deionized water for the same time (Bergomi et al. 2002). For NAA estimation of Cr, Co, Fe, Se, Zn, manganese (Mn), and aluminum (Al), toenails were cleaned in a distillery apparatus with diethyl ether for 30 min, acetone for 40 min, and twice-distilled water for 15 min, thus presenting two washing methods for two different element groups for ICP and NAA analyses.

Horn-Ross et al. (2001) and Chandra Sekhar et al. (2003) scrupulously checked the completeness of washing systems. After washing in deionized water ultrasonically and oven drying at 50°C, toenails were examined using low-power magnification to physically verify the thoroughness of washing. If further washing was required in one or two samples, washing continued again or the samples were excluded altogether from analysis.

D. Desiccation

After washing, the nails are placed in a drying oven at 50 °C for 2 hr, the time required to achieve a constant weight (Rayman et al. 2003). Bergomi et al. (2002) reported two procedures for desiccating toenails: drying at 100 °C for 2 hr for ICP measurement of Cd, Pb, and Cu, and drying at 50 °C for 48 hr for NAA measurement of Cr, Co, Fe, Se, Zn, Mn, and Al. Samples are even desiccated at room temperature for 24 hr because some elements (Hg) are lost at 40 °C. The dried samples, typically 10–135 mg, are weighed into polyethylene capsules for direct analysis with NAA or kept in polyethylene bags for later digestion.

E. Digestion

The dried nail samples are either wet acid digested or dry ashed followed by acid digestion. Rodushkin and Axelsson (2000a) adopted the procedure of digesting 50 mg nails with addition of 0.5 mL HNO_3 and 0.5 mL H_2O_2 in a microwave digestion system. Bergomi et al. (2002) digested 100–200 mg samples with 2.5 mL HNO_3 and 7.5 mL deionized water in a Perkin Elmer microwave digestion system. It is possible to determine directly Zn levels in fingernail samples as small as 20 μg using an AAS furnace. However, where sample size is not a limitation, wet ash digestion before furnace determination is preferred (Sohler et al. 1976). Chen et al. (1999) demonstrated that the arsenic concentration of in-house nail reference material was not substantially different from microwave-digested and room temperature-digested (36 hr) samples. They recommended that the digested samples could be kept in storage for a maximum of 5 d for further analysis, because the samples analyzed for concentrations of elements within this period showed greater quality assurance.

F. Instrumental Measurement

Element levels are estimated in nails using diverse analytical instruments. Although they have their own advantages and disadvantages, one or two instruments are employed for measurement of most elements because each has a specific sensitivity for a particular group of elements. In Table 1 are shown the various instruments employed globally for measuring concentrations of multi-elements in nails. AAS has been predominantly used since 1970, and the others, such as ICP/DCP, ICP-mass spectrometry (MS), photon induced X-ray emission spectroscopy (PIXE), instrumental NAA (INAA), and NAA, have been in use for the past two decades with tremendous improvement in sensitivity of measurement.

G. Quality Control

Monitoring metal bioaccumulation in nails requires precision and accuracy in analysis for decision making in occupational and environmental health (Subramanian and Sukumar 1988; Iyengar and Woittiez 1988). In addition to significant advancement in nail analysis, an improved quality assurance, which is essential, depends on employment of proper certified reference materials (CRM) and participation in the interlaboratory quality control programme (M'Baku and Parr 1982; Parr et al. 1988). Currently, the reference material for toenails is available from National Institute of Standard and Technology (USA) and can be used to ensure the validity of adopted methods. Additionally, an in-house reference material can be prepared from nail samples of healthy volunteers following methods described by Chen et al. (1999).

III. Influencing Factors

Levels of trace elements measured in nails are influenced by various factors that are very specific to a particular subject and region. These factors are discussed based on their distinct and salient characteristics.

A. Age

Age of subjects from whom nails are sampled is found to alter levels of elements in nails. Al and V in nails decreased with age of persons from industrialized and village areas along the bank of the Wogupmeri River in New Guinea (Masironi et al. 1976). Concentrations of Se-TN declined with age in U.S. women (Hunter et al. 1990a). A slight decline in As-TN with age was also observed (Karagas et al. 2000). There was no correlation between age and levels of various elements, particularly Ni in fingernails (Ni-FN) in healthy individuals (Gammelgaard et al. 1991), Cd-, Pb-, and Zn-FN (Hayashi et al. 1993), Se-TN (Kardinaal et al. 1997), Cd-, Cu-, Cr-, Se-, and Fe-TN (Bergomi et al. 2002), Co-, Cr-, Fe-, Na-, and Sb-TN (Rodushkin and Axelsson 2000b), and Se-TN (Swanson et al. 1990). Ultimately, due to age either change or no change in elemental levels is observed, but such a result is not specific to either fingernails or toenails of subjects or to any element.

B. Sex

Concentrations of Pb, Cd, and Zn were higher in nails of males than of females (Nowak and Chmielnicka 2000). In contrast, Pb and Zn in toenails of female

Table 1. Analytical instruments used in nail/element studies.

Subjects	FN/TN/N	Elements	Instruments	Reference
General subjects	FN	Trace elements	AAS	Harrison and Tyree 1971
Exposed subjects in Fairbanks, Alaska	N	As	AAS	Harrington et al. 1978
General subjects from Machakos District, Kenya	N	9 elements	INAA	Othman and Spyrou 1980
Chronically exposed subjects	N	As	AAS	Olguin et al. 1983
Patients with chronic renal failure	FN	Al	AAS/NAA	Marumo et al. 1984
General subjects	FN	As	PIXE	Biswas et al. 1984
Wood treatment factory workers from Nigeria	N	As	AAS	Ndiokwere 1985
Controls and Fe-deficient subjects	FN	S, Ca, Fe, Cu, Zn	XMA	Djaldetti et al. 1987
Subjects of five nations	N	18 elements	ICP	Takagi et al. 1988
General U.S. subjects	FN	17 elements	INAA	Vance et al. 1988
General subjects	N	Ca, Cr, Mn, Fe, Ni, Cu, Zn, Se, Br, Pb	PIXE	Lapatto et al. 1989
Psoriasis patients	N	Ca	EDXMA	Kao 1990
Deceased lead smelter workers and controls		Pb	XFS	Gerhardsson et al. 1993
People of Nigeria	FN	20 elements	INAA	Oluwole et al. 1994
As exposed subjects in West Bengal	N	As	AAS	Das et al. 1995

Table 1. (Continued)

Subjects	FN/TN/N	Elements	Instruments	Reference
Men and women from Prague	TN	12 elements	INAA	Rakovi et al. 1997
Normal subjects	N	As	NAA	Nichols et al. 1998
Urban subjects from Sweden	FN	71 elements	ICP-MS	Rodushkin and Axelsson 2000b
Controls and patients of chronic pulmonary disease	N	Zn	AAS	Leon-Espinosa de los Monteros et al. 2000
Controls and cancer subjects	TN	Se	NAA	Ghadirian et al. 2000
Workers in welding, drilling, or polishing factory.	N	Cr, Fe, Mn, Mo, Ni, V	INAA	Kucera et al. 2001
Controls and subjects of amyotrophic lateral sclerosis (ALS) from Italy	TN	Zn, Se, Cu, Mn, Cr, Cu, Fe, Al, Cd, Pb	ICP-MS, INAA	Bergomi et al. 2002
Controls and coronary heart disease (CHD) subjects from U.S.	TN	Hg, Se, Cd	INAA	Yoshizawa et al. 2002
Controls and prostrate cancer patients from Washington	TN	Cd, Zn	Flame AAS	Platz et al. 2002
Controls and preeclamptic patients from U.K.	TN	Se	INAA	Rayman et al. 2003
Controls and exposed subjects	N	As	ICP-MS, PIXE, GFAAS	Chandra Sekhar et al. 2003
Controls and epileptic patients	N	Mn, Zn, Cu	AAS	Ilhan et al. 2004

TN: toenail; FN: fingernail; N: Nail; AAS: atomic absorption spectroscopy; GFAAS: graphite furnace atomic absorption spectroscopy (AAS); PIXE: proton-induced X-ray emission spectroscopy; NAA: nuclear activation analysis; INAA: instrumental NAA; ICP: inductively coupled plasma spectroscopy; ICP-MS: ICP mass spectroscopy; EDXMA: Energy-dispersive X-ray microanalysis; XFS: X-ray florescence spectroscopy; XMA: X-ray microanalysis.

controls and Pb and Cd in the toenails of female patients with amyotrophic lateral sclerosis (ALS) were greater than those found in male controls and patients (Bergomi et al. 2002).

Many investigations reported no gender-related differences for Hg-FN and Hg-TN (Suzuki et al. 1989), Se-TN (Swanson et al. 1990), Cd-, Pb-, and Zn-FN (Hayashi et al. 1993), Hg-FN in healthy individuals (Gammelgaard et al. 1991), Zn-TN and Cu-TN in Polynesian residents (McKenzie et al. 1978), and Cd-, Pb-, and Hg-N in residents of Mansouria City, Egypt (Mortada et al. 2002).

In nails, Ag, Au, Zn, and Se were higher in females than in males from a nonindustrial U.S. town (Lexington, Kentucky), whereas, Ca, Co, Cr, Hg, and Sc were not different due to sex (Vance et al. 1988).

C. Fish and Alcohol Consumption

Concentration of Hg-FN was correlated with the number of fish meals of subjects from the Province of Rome (Pallotti et al. 1979). Differences in eating habit or in diet caused great variation in Se-FN in southeastern (higher), southwestern (lower), and northern (in between) parts of the Hungarian population (Bogye et al. 1993). Se-TN values of four vegetarians in a preeclamptic group from Oxford did not differ from nonvegetarians (Rayman et al. 2003). Levels of Se-TN (Kardinaal et al. 1997; Hunter et al. 1990a) and Zn-N (Leon-Espinosa de los Monteros et al. 2000) were not affected by alcohol consumption.

D. Smoking

Cigarette smoke is a major source of certain elements (Cd, Pb, and Hg) for the nonoccupationally exposed (Mortada et al. 2004). The concentration of Cd-N was an average of 10 times greater in smokers than in nonsmokers (Roduschkin and Axelsson 2000b). Tobacco smoking with use of the hookah (water pipe) was associated with increased Cd-FN levels in males and females living adjacent to New Delhi (Sukumar and Subramanian 1992a). Persons who currently smoked cigarettes had significantly lower Se-TN (0.553 μg/g) than ex-smokers (0.599 μg/g) and those who had never smoked (0.652 μg/g) (Kardinaal et al. 1997). Similarly, Se-TN was lower in smokers than in nonsmokers from Montreal (Ghadirian et al. 2000), South Dakota, Wyoming (Swanson et al. 1990), and northern Italy (Krogh et al. 2003). The lower level of Se was ascribed either to the nature of tobacco, which reduced Se absorption, or to smokers' consumption of certain foods containing less Se (Ghadirian et al. 2000) or to low Se intake by smokers (Swanson et al. 1990). However, no effect of smoking was found for Cd, Pb, and Hg in residents from Mansouria City, Egypt (Mortada et al. 2002, 2004), Ni-FN in healthy individuals (Gammelgaard et al. 1991), and Zn-N in controls and patients with chronic obstructive pulmonary disease (Leon-Espinosa de los Monteros et al. 2000). Due to smoking, the levels of nonessential elements (Cd and Pb) are higher while the level of an essential element (Se) is either lower or without change.

E. Regional Influence

Alexiou et al. (1980) reported that the petrographical composition of the region influences the concentration of Cu-, Zn-, Fe-, and Mg-FN in normal children. Se-FN and Se-TN were observed to be higher in the southwestern region than in the northern part of the Hungarian population (Bogye et al. 1993). Variations in Se levels due to regional difference were illustrated in the following studies: Se-FN and Se-TN in the population from east Hungary were significantly higher than in the northern people and lower in the southern people of Hungary (Alfthan et al. 1992). Such regional variations among different regions are due to changes in eating habits (Bogye et al. 1993) and food, which ultimately depend on the availability of soil Se for plants (Alfthan et al. 1992). Hunter et al. (1990a) substantiated that the geographical variation in Se-TN was consistent with the geographical distribution of Se in forage crops.

Among 10 centers in Europe and Israel, Se level was the lowest in Germany (Kardinaal et al. 1997). Kasperek et al. (1982) showed that Fe-FN and Se-FN declined from south (Aswan) to north (Alexandria) and Cd-FN was elevated in the Aswan area of Egypt. In an international comparison, Takagi et al. (1988) reported that among 21 elements analyzed in nails of subjects from five nations, certain elements remained high, mainly Ca and Zn in Polish, Hg in Japanese, P and Al in Indians, and Japanese and Cu and Pb in Americans and Canadians.

F. Time and Season

Although fingernails were sampled three times from male and female subjects with an age range of 23–56 yr, total Hg and inorganic Hg were almost consistent for three different formation periods of nails (Suzuki et al. 1989). However, As was slightly greater in the toenails of U.S. control subjects collected during summer months than in other seasons (Karagas et al. 2000). Hence, the time of nail sampling between cases and respective controls becomes a very important factor, stressing that the time gap between two groups for collection must be narrow because the status is also related to the length of exposure and disease (Mannisto et al. 2000).

G. Genetic Variation

By comparing the within-pair correlation for the level of Zn-N in Hungarian adult twin pairs of different zygotes, a very much higher correlation in monozygotes than in dizygotes was observed (Forrai et al. 1984), suggesting that the Zn-N content may be genetically controlled.

H. Nail Character

Levels of Ca, Mg, Al, Cu, Zn, and Fe were not significantly different between normal fingernails of healthy subjects and brittle nails of subjects considered patients because of having brittle nails (Lubach and Wurzinger 1986). A significant variation was found in Ca values between involved and uninvolved nails and between the dystrophic and normal nails of patients with psoriasis (Kao 1990), and levels of As were dissimilar in the various transverse 0.5-mm length of toenail segments (Henke et al. 1982).

I. Contamination

There has been significant environmental As contamination of toenails in exposed and control populations from Australia, because the values of As-TN are more strongly associated with both drinking water and soil As concentration (Hinwood et al. 2003). Extremely high levels of Se-TN found in eastern Finland are usually attributed to contamination by Se-rich shampoo (Mannisto et al. 2000). Zn from an industrial exposure (galvanizing) or from the use of Zn-based creams elevated Zn-TN in normal healthy adult students of New Zealand (McKenzie 1979). Accumulation of cobalt or chromate in the nails has been demonstrated during daily repeated exposure to low cobalt or chromate concentrations by immersing fingers for 10 min (S01) d for 2 wk (Nielsen et al. 2000). The free surface of nails adsorbs a considerable quantity of Ca from the environment, and it is reasoned that such adsorption occurs due to ion exchange; this is evident from the fact that sulfur does not vary over nail cross sections (Forslind et al. 1976). Similar studies are required for nail properties of adsorbing other elements from the environment to determine the possibilities of assessing the extent of contamination.

It is clearly seen that nails are probably contaminated through exposure to air, water, soil, shampoo, nail polish, and creams. In view of this, it is obligatory to consider these factors at the time of sampling and to exclude such samples that would possibly have been contaminated.

J. Other Factors

Hg-FN is correlated with Hg amalgam dental fillings of subjects (Pallotti et al. 1979). Deficiency of nutritional Fe and Zn, which is universally prevalent, causes changes in the characteristics of nails (Weismann and Hoyer 1982; Sato 1991). A patient with thallium (Th) poisoning caused by repeated exposure to low doses of Th had nails with dystrophy in the form of whitish lunular stripes (Saha et al. 2003). An unusual case of yellow nail syndrome was reported in a rare clinical condition (Arroyo and Cohen 1993). Hence, any change in the characteristics of nails related to element status may be viewed as a factor. Unless the factors at maximum levels are not regarded as prerequisite during sampling, analysis and interpretation of data objective analysis cannot be achieved.

IV. Environmental Exposure

To find the sources and possible status of exposure is of prime importance in analyzing nail content of elements in persons exposed to elements through general pathways such as water, food, and air (Barceloux 1999), which receive elemental contaminants mainly from traffic and industrial emissions (man-made) and soil of particular regions (natural source).

A. Drinking Water

Arsenic concentration in nails of individuals who have been exposed through drinking water with As above maximum permissible limit (MPL) of 12.58 μg/L is higher than the value of a control group (4.88 μg/L) and is evident in the cutaneous sign of arsenium in As-exposed general populations from four affected districts in West Bengal, India (Basu et al. 2002). Likewise, the levels of As-TN in residents are explicitly correlated with As levels of drinking water of five regions such as Alaska (Harrington et al. 1978), Mexico (Cebrian et al. 1983), northeast coast of Taiwan (Chiou et al. 1997), New Hampshire, USA (Karagas et al. 2000), and Australia (Hinwood et al. 2003). The value of As-N is elevated (0.5–1.63 mg/kg) in villagers with skin arsenicosis from Patancheru near Hydrabad, Andhra Pradesh (India), who are exposed to As through contaminated water, in comparison to the value of control subjects (0.12–0.16 mg/kg) (Chandra Sekhar et al. 2003). In addition, a 10-fold increase in As-contaminated well water was reflected by a 2-fold increase in As-TN (Karagas et al. 1996).

Drinking water levels of Pb and fluoride (F) are found related to the levels of Pb and F in fingernails of healthy donors from Kuwait (Bu-Olayan et al. 1996) and in the nails of Hungarian children (Schamschula et al. 1988). High nail levels of As, Mn, Pb, and Ni observed in victims are ascribed to As exposure through drinking water (Samanta et al. 2004).

B. Food

Correlation is found between diet and toenail levels of As and Hg in men and women from across the U.S. (MacIntosh et al. 1997), between fish consumption and Hg-TN in U.S. subjects (Yoshizawa et al. 2002) and increased Hg, Al, and Fe in the nails of male and female Japanese (Takagi et al. 1988). Similarly, high concentrations of As-N in villagers with skin arsenicosis and Al-, Fe-, and Sn-N in Indians are ascribed to exposure to As-contaminated milk, vegetables, and other food items, respectively (Chandra Sekhar et al. 2003), and to extensive use of tin-coated iron and aluminum cookware (Takagi et al. 1988). Further, high exposure is observed through nail levels of As, Mn, Pb, and Ni, which may be from the food (Samanta et al. 2004). In the case of iodine (I), however, both negative as well as positive profiles have been found; the correlation between I levels of diet and nails are minimal, suggesting that these biomarkers may not be a good reflection of dietary I (Horn-Ross et al. 2001), whereas Morris et al. (1983) found that nail clippings from the large toe provided an integrated measure of I exposure over a 2- to 4-wk period approximately 1 yr before clippings.

C. Air and Soil

Pb-N in traffic policemen showed significant and positive correlation with duration of exposure of Pb from automobile exhaust (Mortada et al. 2001), whereas concentrations of As-TN in controls and an As-exposed population from Australia were strongly correlated with soil As concentration (Hinwood et al. 2003). Swanson et al. (1990) pointed out that Se-TN indicated elevated Se intake of subjects living in high-Se areas of South Dakota and Wyoming.

D. Urban and Rural Gradients

In multielement analyses, Rodushkin and Axelsson (2000b) observed that among 71 elements analyzed, particularly the fingernail levels of Hg, Cd, Pb, antimony (Sb), and bismuth (Bi) were at elevated profiles indicating higher exposure of an urban population living in the northeast of Sweden than in rural controls. Likewise, Hg-FN levels were linked between urban and rural residents from the Province of Rome (Pallotti et al. 1979). The mean Se-FN of men and women from Se-endemic rural areas of Nawan Shahr District, Punjab, India, was tens of times higher than that of nonendemic rural areas and was associated with clinical symptoms of Se toxicity in some of the endemic subjects (Hira et al. 2004). On the other hand, when chemical symptoms of Se toxicity are not related to Se levels in toenails or blood, there is no evidence of Se toxicity in the subjects residing in western South Dakota and eastern Wyoming, although in these places unusually high natural Se intakes are found (Longnecker et al. 1991). However, high Se levels are observed in nails of subjects from China where chronic Se poisoning through excessive environmental exposure is reported (Barceloux 1999). Natural or man-made sources of element exposure are manifested in the form of high nail levels mainly for As, Pb, and F.

The other major pathways of As exposure of children from arsenic acid-treated wood appear to be oral ingestion by children's hand-to-mouth activity and dermal absorption (Hemond and Solo-Gabriele 2004). It has been recommended that the nail samples of children less than 5 yr of age are the best choice of biological samples for monitoring environmental exposure to elemental sources, in that young subjects are especially prone to increased Cd and Pb exposure, evident from enhanced levels of elements in their nails (Wilhelm et al. 1994). Generally, urban and rural gradients of element exposure are explicitly evident in the nail levels of Pb, Cd, and Hg, which are higher in urbanites than in villagers.

V. Occupational Exposure

With the purpose of predicting human exposure to specific sources of elements from the work place, a number of prospective studies have documented that the greater the levels of trace elements found in nails, the greater the exposure to these elements. When compared to controls, significantly increased profiles have been reported for Sb-N in workers of an antimony refinery (Katayama and Ishida 1987), F-N in workers of a phosphate fertilizer plant (Czarnowski and Krechniak 1990), Cr-N in workers of a hydraulic production plant in Teneu, Town of Yambol, Bulgaria (Madzhunov et al. 1993), Zn-N, Cu-N, and Cd-N in radiographers (Majumdar et al. 1999), Ni-FN in two groups of occupationally exposed workers (Peters et al. 1991), Hg-TN in U.S. dentists (Yoshizawa et al. 2002), and Hg-FN and Hg-TN in dentists from western Scotland (Ritchie et al. 2002). In addition, a high correlation coefficient has been obtained in a comparison of mean air As concentrations of high-, medium-, and low-exposure groups of workers with corresponding As level in their fingernails (Agahian et al. 1990).

A. Low Levels with Limited Exposure

Hewitt et al. (1995) found mean levels of As-FN well within normal values and that As exposure was not apparent among employees of arsenical pesticide-producing industries.

B. Nail Levels vs. Exposure Period

Mortada et al. (2001) reported that the level of Pb-N in traffic policemen showed excellent and positive correlation with duration of exposure to Pb as duration of employment and concluded that increased levels of Pb-N were ascribed to extended period of exposure. Gerhardsson et al. (1995) found that Pb-N levels were the same in both controls and deceased long-term exposed male lead smelter workers.

C. Occupational Exposure and Health Risk

Tsolova et al. (1995) found that occupational groups of a copper smelter in Pirdrop, Bulgaria, had both high and low levels of As-N and proposed that degree of exposure corresponded to low health risk and also that some of the most heavily exposed occupational group might be expected to reach serious levels of intake and health risk. Feldman et al. (1979) reported that values of As-N were lower in controls than in workers exposed to arsenic trioxide in a copper-smelting factory with clinical and subclinical symptoms of neuropathy.

Thus, for the majority of elements (Cr, F, Zn, Cu, Cd, Hg, Sb, and As), apparent exposure status is higher in the workers of various factories than in controls. In the case of Pb-N, the results are inconsistent (Gerhardsson et al. 1995; Mortada et al. 2001).

VI. Health Status Effects

Element levels measured in nails are evaluated for predicting the role of elements in etiology, implication, or complications of disease, and hence it is proposed that nails may be considered as a diagnostic tool for different ailments related to body storage of elements.

A. Nutritional Elements

A majority of investigations have advocated the use of fingernails or toenails for characterizing the relationship of health status with the concentrations of Cu in patients with Wilson's syndrome (Martin 1964) and in patients with cystic fibrosis (CF) (Fite et al. 1972), Ca and Mg in chronic uremia patients (Lim et al. 1972; Robson and Brooks 1974), and Se in subjects with dental caries (Hadjimakos and Sheaver 1973). In the case of Se, three patterns have been reported in nails of patients. First, a uniformly positive status, i.e., a statistically significant inverse association, was found between the levels of Se-TN and the risk of colon cancer in both genders from Montreal (Ghadirian et al. 2000), lung cancer (van den Brandt et al. 1993a), prostrate cancer (Yoshizawa et al. 1998; van den Brandt et al. 2003), bladder cancer (Zeegers et al. 2002), and myocardial infarction (Kardinaal et al. 1997). In a case-control study evaluating the association between Se-TN and lung cancer risk in male smokers, a protective association for high Se status among men and low Se-TN status may be related to increased risk for lung cancer (Hartman et al. 2002). Levels of Se-TN were significantly lower in pregnant preeclamptic subjects (Rayman et al. 2003) and in patients with acute myocardial infarction (AMI) (Kok et al. 1989) than their matched controls. In both cases low Se status is associated with serious disease conditions.

Second, a negative trend, i.e., no substantial relationship between concentrations of Se-TN and decreased risk of breast cancer has been depicted in cases from regions of the U.S. (Hunter et al. 1990b), the Netherlands (van den Brandt et al. 1994), five European countries (van't Veer et al. 1990), and Finland (Mannisto et al. 2000). Similarly, levels of Se-TN do not indicate the risk of other diseases such as nonfatal AMI studied in 10 centers from Europe and Israel (Kardinaal et al. 1997), cancer (uterine, colorectal, melanoma, ovarian, lung) cases from 11 states of the U.S. (Garland et al. 1995), breast cancer from the Netherlands (van Noord et al. 1987; van't Veer et al. 1990), coronary heart disease (CHD) (Yoshizawa et al. 2003), and prostrate cancer risk (Allen et al. 2004).

Third, a mixed trend of association has been pointed out in the following studies: men with oral cancer from western Washington state had lower level Se-TN than the controls, whereas women with oral cancer did not have any such change in Se level (Rogers et al. 1991). A statistically significant inverse relationship was found between Se-TN and the risk of colon cancer for both genders combined and for subjects from Montreal, and also no significant connection was discerned between Se-TN and breast or prostrate cancer (Ghadirian et al. 2000). Further, an inverse relationship was found between Se-TN levels and stomach cancer in men but not in women with stomach cancer and subjects with colorectal cancer (van den Brandt et al. 1993b).

Similar to Se, other nutritional element values have been connected to ill health of subjects in a few reports. A marginally significant positive association has been observed between Zn-FN and CHD and hypertension (Sukumar and Subramanian 1992b), Fe-N, Ca-N, and Co-N and esophageal cancer cases (Rogers et al. 1993), Cr-TN and breast cancer risk among postmenopausal U.S. women (Garland et al. 1996), and Zn-FN and osteoporosis in elderly Japanese women (Karita and Takano 1994). The amount of Fe in nail samples reflected the Fe status of healthy individuals, iron-deficient subjects before and after dietary iron therapy, and four patients at various stages of iron treatment (Sobolewski et al. 1978).

Other studies have contradicted that fingernails and toenails do not exhibit differences in element levels of controls and patients, particularly, levels of Cu in children with CF (van Stekelenburg et al. 1975), N-FN in intrauterine growth-retarded neonates (Brans and Ortega 1978), Ca-N in subjects with psoriasis (Kao 1990), Zn-N in patients with chronic obstructive pulmonary disease (Leon-Espinosa de los Monteros et al. 2000), I-N in a high- and low-risk papillary thyroid cancer group residing in the San Francisco Bay area (Horn-Ross et al. 2001), Zn-TN in pellagara patients (Vannucchi et al. 1995) and in subjects with AMI from eight European countries and Israel (Martin-Moreno et al. 2003).

B. Toxic Elements

Generally, high levels are found when toxic or nonessential elements are measured in nails of patients. Narang et al. (1987) reported that the level of As-N was significantly higher in Indian opium addicts than in controls. Shrestha et al. (2003) showed that the accumulation of As in nails was higher than the acceptable level and that it could be linked with the prevalence of dermatosis-related arsenicosis in the Terai region of Nepal population exposed to As contamination through groundwater (shallow tube wells).

Similarly, Karagas et al. (2001) confirmed that the values of As-TN might provide an index or biomarker for reliable estimates of internal As exposure that could be used to assess cancer risk. Nickel- sensitive women had significantly higher level of Ni-TN than control subjects, whereas there was no significant difference in Ni-FN concentrations between the two groups (Gammelgaard and Veien 1990).

A weak correlation was reported between the Hg-TN and CHD (Yoshizawa et al. 2002), unlike the direct association pointed out between the Hg-TN and the risk of MI in men from eight European countries and Israel (Guallar et al. 2002). Dental fluorosis in Hungarian children was directly related to the level of F-N (Schamschula et al. 1988). Further, the highest value of F-TN was associated with intake and risk of osteoporotic fracture, specifically hip facture among women (Feskanich et al. 1998), whereas no association was found between the Hg-N and changes in the psychomotor response of dentists from western Scotland (Ritchie et al. 2002).

C. Multielements

Compared with controls, patients with CF had increased levels of sodium (Na), potassium (K), and chlorine (Cl) (Roomans et al. 1978) and Na and K (Kapito and Shwachman 1964; Kapito et al. 1965) in their nails. Children with CF had a significant increase of Cu and Zn levels in nails but carriers did not (Escobar et al. 1980).

Karita et al. (2001) observed that levels of Zn-FN, but not Ca-FN and Mg-FN, were significantly higher in patients with osteogenesis imperfecta (OI) than in normal subjects and suggested that levels of Zn-FN may reflect abnormal Zn metabolism in OI and osteoporosis. Bergomi et al. (2002) found an inverse relation with values of Se and Zn, a direct correlation with Cu level, and also no evidence of an association between levels of Cd, Pb, Cr, Co, Fe, and Al in the toenails of patients with risk of ALS from the Emilia Romagna region in northern Italy.

Rogers et al. (1993) presented diverse patterns of nail levels of elements for various cancer patients, mainly low Fe in larynx and esophageal cancer, enhanced levels of Fe and Ca in esophageal cancer, and elevated Co in esophageal and oral cancer patients. They concluded that there may be a difference in mineral intake or metabolism between individuals who develop carcinomas of the upper aerodigestive tract and those who do not.

Hepatosplenic subjects had both elevated levels of Cl and I and decreased levels of Mg, Ca, K, Mn, Cu, and Se in fingernails compared to controls from rural areas east of Alexandria, Egypt (el-Khatib and el-Mohandes 1992). Compared with matched controls, a statistically significant decrease in the percentage of Cu, Fe, and Zn and an increase in sulfur (S) and no difference in Al were found in fingernails of patients with chronic renal failure treated by hemodialysis (Zevin et al. 1991).

The concentration of Zn-N was significantly lower in female subjects with CHD and hypertension than that of controls from New Delhi (Sukumar and Subramanian 1992b). Similarly, the levels of Cr, Fe, and Zn in fingernails of aged patients with hypertension and CHD were significantly lower than of aged healthy controls (Tang et al. 2003). An increased Mn level was detected in nails of patients with epilepsy compared to controls, whereas the concentration of Zn-N was found to be unchanged (Ilhan et al. 2004).

When levels of bromine (Br), Hg, K, and Zn were determined by INAA in nails of patients with Alzheimer's disease (AD) at 6-mon intervals for up to 3 yr, no change for Br, a decreasing trend for Hg, and an increase for K and Zn with increasing age of patients were reported. Accordingly, element imbalance was shown in nails of AD patients (Vance et al. 1990).

Many studies pointed out that nail levels of elements were not significantly different between certain patients and controls, particularly with Cd-TN and Zn-TN in prostrate cancer cases from Washington state (Platz et al. 2002), S, Cd, Fe, Cu, and Zn in fingernails of women with iron deficiency (Djaldetti et al. 1987), and Cu-TN and Zn-TN in Polynesian men and women in different ranks of systolic blood pressure (McKenzie et al. 1978).

VII. Element Interaction

Exposure to multielements results in interaction in the human body, which is noticeable in the form of additive effects of toxic elements (Hg, Cd, As, etc.) and antagonistic effects of nutritional (Se, Cr, Cu, Zn, Mg, etc.,) elements against toxic elements. Such interaction is also illustrated in various tissues by comparing elemental levels between toxic or nutritional elements. Presently, there is a major thrust to learn the extent nails reveal the interaction of toxic and nutritional elements in human beings.

Nowak and Chmielnicka (2000) found an increase of Pb-FN resulted in decreased FN levels of Zn and Cu, and also similarly in blood and hair, level of Pb-FN significantly influenced Zn and Cu levels, which are essential for proper body function, thus revealing the antagonistic activity of Pb against the essential nutrients. In contrast, Vance et al. (1990) reported a decrease of Hg and increase of K and Zn when levels of K, Br, Zn, Cu, and Hg were determined in the nails of patients with AD. Thus, a pattern of antagonistic effects has been shown in the case of elevated nutritional elements (Zn and Se) causing a decrease of Hg (Vance et al. 1990). Among 10 elements (As, Hg, Pb, Ni, Cd, Mn, Se, Zn, Cu, and Fe) analyzed, Mn and Ni levels alone were correlated with those of essential elements (Zn, Cu, and Fe), thereby indicating that Mn and Ni may substitute for those elements in nails of arsenic victims from an arsenic-affected area of West Bengal, India (Samanta et al. 2004).

No interaction was found between Se-TN and antioxidant intake from food by breast cancer patients and controls from Finland (Mannisto et al. 2000), and between Hg-TN and in antagonistic effect the low- and medium-Se categories with the risk of CHD (Yoshizawa et al. 2002).

VIII. Element Speciation

The species or state of elements determines their bioavailability for body uptake and, in turn, affects their levels in nails (Barceloux 1999). Moreover, the different oxidation states of nail elements are of specific interest, owing to different levels of toxicity or physiological roles exhibited. At present, a large majority of investigations have given more emphasis to analyzing nail levels of total inorganic elements rather than to their speciation. Mandal et al (2003) observed that fingernails and hair samples that were collected from the As-affected area of West Bengal, India, contained As (III), As (V), monomethylarsenic acid, MMA (V), and dimethylarsenic acid, DMA (V), with some containing only dimethylarsonous acid, DMA (III); they concluded that DMA (III) and DMA (V) content in fingernails could serve as biomarkers to As exposure.

IX. Influence of Supplements

The validity of element levels in nails as a measure of element intake can be assessed based on the studies carried out with element supplementation and its subsequent measure in nails. The mean Se-TN was higher in two groups of U.S. female consumers of Se supplement than in nonusers, and the dose–response relationship in two groups was also significant. The concentration of Se-TN was shown to reflect Se dietary intake (Hunter et al. 1990a). Ovaskainen et al. (1993) also found that Se-TN concentrations of urban men aged 55–60 yr were the best predictors of Se intake from supplements. In another study, a reasonable association was observed between I supplementation and levels of I-N in Malawi children and between high I consumption and levels of I-N in Utah men

(Spate et al. 1995). So far, the supplementation of Se and I and nail levels have been interrelated, but for other elements further studies are required.

In a 1-yr trial studying the effect of high- selenium bread on Se concentrations in blood and toenails, Longnecker et al. (1993) found Se-TN concentration was unaffected by dietary intake the previous 3 mon and appeared to provide a time-integrated measure of intake over a period of 26–52 wk. Ultimately, they found that use of Se concentration in toenail clippings might be an alternative to blood level when a measure of long-term average intake was desired. The absence of short-term effect of diet on Se-TN concentration also makes nails a useful biomarker of intake in retrospective studies.

X. Correlation of Element Levels

Estimation of multielements at a given time in the nails is one of the best-known advantages of regarding it as an indicator of element status. Because multielements have different biological implications, many investigations have presented the relative levels of varied elements. Among various elements analyzed, a significant correlation was observed between levels of Co and tungsten (W) in toenails of subjects occupationally exposed to hard metal dust (Sabbioni et al. 1994), Cd, Cu, Pb, and Zn in toenails of children 3–7 yr of age living in an industrialized area and in a rural area of Germany (Wilhelm et al. 1991), Al and vanadium (V) in the toenails of a population from New Guinea (Masironi et al. 1976), and Ca and P in the fingernails of elderly Japanese women (Karita and Takano 1994). Such studies reporting positive correlations between elements are meager; the number of elements between which significant correlation found is also extremely rare.

XI. Relative Element Levels in Nails and Other Samples

Comparison of nail levels of elements is made with that of other tissues such as hair, blood, plasma, urine, and teeth collected from the same subjects. Because elements are either nutritional or toxic, they are involved in metabolic activities and are accumulated in various tissues at different concentrations. With this perspective, the relative levels of elements in nails and other tissues or samples are indicative of physiological changes, body storage of elements, the proportionate levels of elements in nails and other tissues, the key tissue as the best indicator or biomarker of a particular element, implication of health status, and status and duration of exposure.

The primary consideration in employing nails for monitoring of elements is not how easily the samples can be collected nor even how accurately trace elements can be determined in nails, but rather what does the content indicate, i.e., to what extent does the measured level reflect the concentration and activity of elements in other parts of the body (Hopps 1977). Table 2 indicates the various reports of relative levels of elements in nails and other samples, and such levels are discussed differently.

Table 2. Correlation of element levels between nails and other samples.

Comparison				
FN, TN, N	Other samples	Elements	Results	Reference
FN	TN	Se	Similar pattern in both samples	Alfthan et al. 1992
FN	TN	Se	Similar regional difference	Bogye et al. 1993
FN	TN	Hg	Similar level in both samples from dentists from western Scotland	Ritchie et al. 2002
FN	TN	Cu	Higher in FN than in TN of children with CF	van Stekelenburg et al. 1975
FN	TN	Total and inorganic Hg	Higher in FN than in TN, indicating FN contamination	Suzuki et al. 1989
FN	TN	Ni	Higher in FN than in TN of Ni hypertensives and controls.	Gammelgaard and Veien 1990
FN	SH	Multielements	No correlation	Kasperek et al. 1982
FN	SH	Be, B, Y, Mo, Ru, Ag, Ra, Cl, Ca, Ge	No correlation	Kasperek et al. 1982
FN	SH	Hg, Cd, Pb, Zn, Hf, Sb, Bi	Highest correlation found	Madzhunov et al. 1993
FN	SH	Hg, Cd, Pb, Sb, Bi	Strong positive correlation	Rodushkin and Axelsson 2000b
FN	Hair	12 elements	Higher in FN than in hair	Oluwole et al. 1990
TN	SH	Cd, Pb	Similar in both samples of children of exposed fathers	Wilhelm et al. 1994
TN	SH	Cu, Zn	Similar in both samples	Wilhelm et al. 1991
TN	SH	Cd, Pb	Higher in TN than SH	Wilhelm et al. 1991
N	SH	Total Hg	Higher in SH than in nails	Suzuki et al. 1989
N	Hair	Essential elements	No correlation	Vance et al. 1988

Table 2. (Continued)

Comparison				
FN, TN,N	Other samples	Elements	Results	Reference
N	Hair	Al, Cl, I, Mg, Ca, K, Cu, Se	Higher in FN than in hair	el-Khatib and el-Mohandes 1992
N	Hair	Nonessential elements	Positive correlation	Vance et al. 1988
FN	Serum	Al, Fe, Zn	No correlation	Suzuki et al. 1989
FN	Serum	Cu	Similar in both samples	Zevin et al. 1991
TN	Blood	Se	Similar (low) in both samples	Longnecker et al. 1991
TN	Urine	As	Significantly correlated	Karagas et al. 2001
N	Urine	Cr	No correlation	Madzhunov et al. 1993
FN	Urine, Hair	As	Similar in all three	Lin et al. 1978
FN	TN, sera	Se	Similar in all three	Bogye et al. 1993
FN	TN & hair	Hg	Similar (higher) in all three samples of dentists than control samples	Ritchie et al. 2002
N	Blood, urine	As	Similar (higher) in all samples of exposed than in controls	Basu et al. 2002
N	Hair, Blood	Pb	Similar in all three	Mortada et al. 2001
N	Hair, urine, sera	As	Similar (higher) in all the four samples of opium addicts	Narang et al. 1987
FN	Muscle	N		
FN	Sera, hemoglobin	Fe	No correlation	Djaldetti et al. 1987
N	Hair, plasma	Ni	No correlation	Gammelgaard and Veien 1990
N	Urine, hair	F	Positive correlation	Czarnowski and Krechniak 1990
TN	Urine, pubic hair	Co, Ta & W	Nails could represent a valuable indicator of exposure	Sabbioni et al. 1994

FN: fingernails; TN: toenails; N: nails; SH: scalp hair.

A. Fingernails vs. Toenails

When fingernails and toenails are compared for elemental levels, three trends are observed. First, similar patterns have been noticed for levels of Se in Hungarians (Alfthan et al. 1992; Bogye et al. 1993). Another study indicates that Cd is higher in toenails than in fingernails (Zhang et al. 2001). Other reports indicate that Ni is higher in fingernails than in toenails (Gammelgaard and Veien 1990). The higher levels of inorganic and total Hg (IHg and THg) found in fingernails than in toenails are ascribed to the possibility of external contamination with IHg in fingernails (Suzuki et al. 1989). This comparison explains that there is likelihood of contamination of nails from environmental sources of Hg and Ni, which may be more obvious in fingernails than in toenails because chances for contact and other means of external contamination are greater for fingernails than for toenails. Hence, toenails are preferable to fingernails in element analysis when there is likelihood of external incorporation of elements. In the case of nutritional elements, fingernails in patients with CF contained higher Cu than did the toenails. Hence, for assessment of essential elements when contamination is not likely, fingernails are most suitable (van Stekelenburg et al. 1975).

B. Nails vs. Hair

Comparison of elemental levels between nails and hair shows the following six types of relationship:

1. No correlation existed between them for levels of Ni (Gammelgaard and Veien 1990).
2. A positive correlation was obtained between them for levels of F (Czarnowski and Krechniak 1990).
3. Similarity has been reported in toenail and scalp hair levels of Cd and Pb in children of exposed fathers (Wilhelm et al. 1994).
4. When compared to hair levels, fingernail levels are at high profile mainly for Se, Hg, Cr, Fe, Zn, Co, Cu, Br, As, Sb, Na, and Sc (Oluwole et al. 1990), for Al, Cl, I, Mg, Ca, K, Cu, and Se (el-Khatib and el-Mohandes 1992), and for Cr, Cs, Fe, Sb, and Sc (Kasperek et al. 1982).
5. Lower levels of total Hg were found in nails than in scalp hair (Suzuki et al. 1989).
6. Mixed trends, i.e., both positive and negative correlations (Rodushkin and Axelsson 2000b), similar status in toenails and scalp hair and higher levels in toenails than in scalp hair (Wilhelm et al. 1991) have been reported. High levels of elements observed in nails may indicate two possibilities; (1) the nail sample may be the best indicator showing maximum levels due to high bioaccumulation or (2) high levels may be due to external contamination.

A correlation has been observed for 17 elements between hair and nails, and with few exceptions, concentrations of nonessential trace elements (Ag, As, Au, Hg, and Sb) are positively correlated in hair and nails, whereas concentrations of essential elements (Br, Ca, Co, Cr, Na, and Se) show no correlation (Vance et al.

1988). Overall, it could be recommended that nails are as good as hair and even better in specific cases.

C. Nails vs. Urine

Three patterns of element concentrations can be derived from studies describing the comparison of element levels between nails and urine.

1. No correlation for the levels of Cr (Madzhunov et al. 1993).
2. Significant correlations for As (Karagas et al. 2001) and F (Czarnowski and Krechniak 1990) are two contradicting results revealing that correlation is significant for nonessential elements such as As and F, which are at excess level, but not for an essential element such Cr, which is deficient.
3. Same status has been observed between nails and urine for the levels of Ta and W (Sabbioni et al. 1994) and As (Lin et al. 1978; Narang et al. 1987; Basu et al. 2002).

D. Nails vs. Blood

1. No correlation is observed between levels of Al, Fe, and Zn in fingernails and serum (Zevin et al. 1991), Fe in fingernail and serum (Djaldetti et al. 1987), and Ni in nails and plasma (Gammelgaard and Veien 1990).
2. Similar levels of Cu in fingernails and serum (Zevin et al. 1991), Se in toenails and blood (Longnecker et al. 1991), Pb in nails and blood (Mortada et al. 2001), Se in toenails and serum (Bogye et al. 1993), and As in nails and serum (Narang et al. 1987) are reported.

E. Nails vs. Multiple Tissues

In comparison between the nails and more than one other tissues, four patterns were observed.

1. No correlation for the levels of Fe in fingernails, serum, and hemoglobin (Djaldetti et al. 1987) and for the levels of Ni in nails, hair, and plasma (Gammelgaard and Veien 1990).
2. A positive correlation for the levels of F in nails, urine, and hair (Czarnowski and Krechniak 1990) and for levels of Pb in nails, hair, liver, and kidney of deceased lead smelter workers (Gerhardsson et al. 1995).
3. Similar levels of Hg in fingernails, toenails, and hair (Ritchie et al. 2002), As in fingernails, urine, and hair (Lin et al. 1978), As in nails, blood, and urine (Basu et al. 2002) and in nails, hair, urine, and serum (Narang et al. 1987), and Se in fingernails, toenails, and serum (Bogye et al. 1993) were seen.
4. Pb-N concentrations were higher than in soft tissues (muscle, heart, aorta, spleen, lung, and prostrate) but lower than in bone Pb (Barry 1975).

Negative as well as positive correlations obtained between elemental levels of nails and other samples are few. In comparing fingernails with toenails, the higher element levels of environmental sources observed in fingernails than in toenails is attributed to contamination because the fingernails have more chances of contact than toenails. In the case of deficient nutritional elements (Fe), the correlation between nails and blood is negative, indicating a reduced relationship of the body store and the observed element level. When excess levels of nonessential elements are compared between nails and other samples, the correlation is significant or the levels are similar in all samples, and thus the utility of nails as a biomarker is as good as that of other tissues such as hair, urine, and blood.

XII. Conclusions

Nails can accumulate not only toxic metals (Pb, Cd, Hg, As, etc.) but also essential elements such as Zn, Cr, Fe, Ni, Co, Cu, Mn, Ca, Na, and K (Nowak and Chmielnicka 2000). When nail levels of elements are compared between controls and subjects of environmentally exposed groups, a positive correlation is obtained for Pb in traffic policemen (Mortada et al. 2001), Hg, Cd, Pb, Sb, and Bi in urban population from Sweden (Rodushkin and Axelsson 2000b), Hg-FN in urban residents from the province of Rome (Pallotti et al. 1979), and Se-FN in Se-endemic rural residents of Punjab, India (Hira et al. 2004).

Levels of As, Pb, F, and Hg are compared in the nails of exposed subjects and sources such as food (Chandra Sekhar et al. 2003), water (Hinwood et al. 2003; Karagas et al. 2000; Bu-Olayan et al. 1996), and soil (Swanson et al. 1990), and possible routes from these sources have been found for bioaccumulation in human subjects.

Similarly, in comparison to controls, nails of workers showed increased profiles for Sb, F, Cr, Zn, Cu, Cd, Hg, As, and Pb (Katayama and Ishida 1987; Czarnowski and Krechniak 1990; Madzhunov et al. 1993; Majumdar et al. 1999; Ritchie et al. 2002; Agahian et al. 1990). Nail levels of elements were related significantly to the exposure period and health risk of workers (Feldman et al. 1979; Tsolova et al. 1995; Mortada et al. 2001). Use of nails in biomonitoring of environmental and occupational exposure of elements has been well documented as either toxic or nutritional elements are found above normal levels.

The outcome of health risk assessment as a consequence of exposure to elements in different patients and controls has revealed either significant or inconclusive results. When results are positively related between controls and patients, Hartman et al. (2002) reported a protective association for high Se among men, and thus low Se-TN may be related to increased risk of lung cancer. As a negative relationship is observed in certain prospective studies, interpretation varies depending upon the causal factors. Nail levels of elements do not indicate disease because the result of investigation is negative (Mannisto et al. 2000; Kardinaal et al. 1997).

In other studies, a diverse pattern of nail levels has been presented for various cancer (Rogers et al. 1993) and ALS patients (Bergomi et al. 2002). Such results in many studies may have various causes. In spite of significant findings that Cu and Zn levels were higher in children with CF than controls, nails could not be used routinely because of the large fluctuations produced by weight of nails, the sensitivity of the resonance line, and variation in these element concentrations on the nail surface (Escobar et al. 1980). Each clipping of nails indicates several weeks of growth and hence nails from controls and patients differ in the time between formation and clipping (Hinwood et al. 2003).

Matching the control to each case on age, sex, smoking status, date of nail collection, and other factors is an imperative criterion for finding association between nail levels of elements and risk of illness. When Yoshizawa et al. (2003) matched control to CHD cases on various factors, Se-TN levels were not associated with the risk of CHD but were inversely associated with risk of nonfatal myocardial infarction. Thus, it is understood that proper interpretation of data with use of various factors is one of most important approaches of nail analysis.

In utilizing toenails as a biomarker to assess exposure in epidemiological studies on cancer and other chronic diseases, random within-person variability in exposure leads to attenuation of association between exposure and disease (Garland et al. 1994). In such case, reevaluation of the individual sample is essential for an extended period. If such an individual sample has reproducibility for subsequent analysis, the toenail concentrations of certain trace elements are a useful biomarker of exposure in which a single sample is assumed to represent long-term exposure.

Due to physiological roles of elements, two classes of abnormalities associated with their levels are (i) a specific deficiency of elements arising from inadequacies or imbalance and (ii) a conditional deficiency resulting from accumulation of innate or toxic elements from environmental exposure, which can displace essential minerals from their metabolically active site and cause them to act directly as a cellular toxin. The various reports using nails for assessing the specific deficiency of elements provide both positive and negative views.

The measurement of Zn-TN in a New Zealand population (McKenzie 1979) and Fe-FN in Fe-deficient patients (Djaldetti et al. 1987) did not reflect their specific deficiency. However, Bogye et al. (1993) reported that Se deficiency could be verified with the use of Se-TN and Se-FN. Kok et al. (1989) reported that Se-TN levels reflected a low Se level that was present before AMI and might be of etiologic relevance. Further, Sobolewski et al. (1978) demonstrated that the Fe status of healthy laboratory staff and Fe-deficient subjects before and after Fe therapy and postmortem cases was reflected by the amount of Fe present in their nail samples. Hence, they proposed that nails were an economical and noninvasive tool for assessing element status of individuals.

Chronic deficiency of Zn and Cu was reported owing to interaction of Pb, which was evident from fingernail levels of high Pb and low Zn and Cu (Nowak and Chmielnicka 2000). Thus, an increase of Pb caused decrease of essential elements (Zn and Cu). Similarly, Vance et al. (1990) reported that high levels of an essential element (Zn) decreased Hg level, which was observed from their levels in nails; however, interaction was not found between nail levels of Se and Hg by Mannisto et al. (2000) and Yoshizawa et al. (2002). At the same time, studies relating to deficiency or baseline levels of various essential elements

(I, Zn, Cu, Cr, Se, Mn, Fe, Mo, F, etc.) are not available, so that determining deficiencies of subjects may be very difficult.

When levels of elements are compared between nails and other samples of the same subjects, the following aspects of relationship are expected to obtain: (a) the key sample indicating the highest level of any specific element, and (b) the order of relationship.

In view of the chemical properties of elements with their metabolic roles and the biological characteristics of tissues, normally it has been observed that, similar to hair, nails can accumulate Hg and As because both have keratin and are rich in sulfur content. The majority of studies have found that Hg and As are accumulated at levels pinpointing exposure. Karpas (2001) has suggested that with the advent of efficient and sensitive methods, use of nails for determinations of uranium (U) can avoid the drawbacks of urine analyses, which may be misleading if urine samples are not collected within a relatively short time after exposure or if the samples are not representative due to collection logistics.

It would be exceptionally difficult to derive the order of relationship in various element levels between nails and other samples because of the heterogeneous nature of elements and samples. Even then, based on the comparison made, there are attempts at a certain level to show that nails are comparable to hair, to some extent to blood and teeth, and possibly to urine. The fact that there are statistically significant relationships between hair and nail concentration with many toxic elements argues for the use of nails along with hair for exposure assessment (Rodushkin and Axelssson 2000b). The available literature is exceedingly meager for relative assessment of nail levels of elements with that of other samples such as milk, saliva, tears, sweat, exhaled air, biliary and fecal materials, and soft tissues from biopsy and autopsy.

In a number of research fields, nail levels of elements have been measured with other relevant parameters to make further use of them for multiple benefits. Longnecker et al. (1996) proposed formulas to estimate the intake of individuals, based on Se levels in a single specimen of toenail of an adult living in south Dakota and Wyoming and multiple iodine duplicate plate food composites and found that the concentration of a single toenail specimen could provide a rough estimate of intake of an element for each subject. In another study, Vecht-Hart et al. (1995) concluded that Ca and Mg measurement in toenails of women from Utrecht, the Netherlands, by INAA could not be used for screening purposes in the prevention of osteoporosis because the correlations between Ca and Mg measurement and bone mineral densities were very low.

Summary

Human nails are extensively employed for monitoring exposure to excessive levels of elements. Nails can be studied easily and economically in subjects of residential areas and industrial workers and patients, and sometimes may indicate that high concentrations of elements are related to various illnesses. Deficiency of certain elements (Se and Fe) can also be determined with use of nails of subjects with specific or unknown deficiencies. Although nails are different from other biological samples in bioaccumulation, physiological activities may be partially due to chemical properties of the elements. Comparison of multielements between nails and other samples is relevant and found significant, because of the status of chronic exposure to high levels of elements. However, comparison of multielements between nails and other samples results in inconsistencies when exposure to elements is acute or unpredictable, or the subject is deficient in the elements being measured. Studies associated with elemental speciation, supplementation, interaction, and deficiency of essential elements (Zn, Cu, Mn, Mo, etc.), although rarely available, may enhance the growing use of nails. In many other research fields nail levels of elements and other associated factors or parameters are being investigated for further expansion of their application.

Acknowledgments

The author is grateful to Professor R. Subramanian, Director, Bio-informatics Center for Medicinal Plants, Chennai, India, and Professor A.L.N. Sharma, Head, Department of Education in Science and Mathematics, Regional Institute of Education (R.I.E.), for their critical suggestions while completing the review, and to Professor G. Ravindra, Principal, R.I.E. (National Council of Educational Research and Training), Mysore, India, for encouragement and the necessary facilities provided.

References

Agahian B, Lee JS, Nelson JH, Johns RE (1990) Arsenic levels in fingernails as a biological indicator of exposure to arsenic. Am Ind Hyg Assoc J 51(12):646–651.

Alexiou D, Koutselinis A, Manolidis C, Boukis D, Papadatos J, Papadatos C (1980) The content of trace elements (Cu, Zn, Fe, Mg) in fingernails of children. Dermatologica 160(6):380–382.

Alfthan G, Bogye G, Aro A, Feher J (1992) The human selenium status in Hungary. J Trace Elem Electrol Health Dis 6(4):233–238.

Allen NE, Morris JS, Ngwenyama RA, Key TJ (2004) A case–control study of selenium in nails and prostate cancer risk in British men. Br J Cancer 90(7):1392–1396.

Arroyo JF, Cohen ML (1993) Improvement of yellow nail syndrome with oral zinc supplementation. Clin Exp Dermatol 18(1):62–64.

Bank HL, Robson J, Bigelow JB, Morrison J, Spell LH, Kantor R (1981) Preparation of fingernails for trace element analysis. Clin Chim Acta 116(2):179–190.

Barceloux DG (1999) Selenium. J Toxicol Clin Toxicol 37(2):145–172.

Barry PS (1975) A comparison of concentration of lead in human tissues. Br J Ind Med 32(2):119–139.

Basu A, Mahata J, Roy AK, Sarkar JN, Poddar G, Nandy AK, Sarkar PK, Dutta PK, Banerjee A, Das M, Ray K, Roychaudhury S, Natarajan AT, Nilsson R, Giri AK (2002) Enhanced frequency of micronuclei in individuals exposed to arsenic through drinking water in West Bengal, India. Mutat Res 516(1–2):29–40.

Bate LC, Dyer FF (1965) Trace elements in human hair. Nucleonics 10:74–81.

Bergomi M, Vinceti M, Nacci G, Pietrini V, Bratter P, Alber D, Ferrari A, Vescovi L, Guidetti D, Sola P, Malagu S, Aramini C, Vivoli G (2002) Environmental exposure to trace elements and risk of amyotrophic lateral sclerosis: a population-based case-control study. Environ Res 89(2):116–123.
Biswas S, Abdullah M, Akhter A, Tarafdar S, Khaliquzzaman M, Khan A (1984) Trace elements in human fingernails: measurement by proton-induced X-ray exission. J Radioanal Nucl Chem 82:111–124.
Bogye G, Feher J, Georg A, Antti (1993) Complex study of selenium levels in healthy subjects in Hungary. Orv Hetil 134(47):2585–2588.
Brans YW, Ortega P (1978) Perinatal nitrogen accretion in muscles and fingernails. Pediatr Res 12(8):849–852.
Bu-Olayan AH, Al-Yakoob SN, Alhazeem S (1996) Lead in drinking water from water coolers and in fingernails from subjects in Kuwait City, Kuwait. Sci Total Environ 181(3):209–214.
Caroli S, Senofonte O, Violante N, Fornarelli L, Powar A (1992) Assessment of reference values for elements in hair of urban normal subjects. Michrochem J 46:174–183.
Caroli S, Alimonti A, Coni E, Petrucci F, Senofonte N, Violante N (1994) The assessment of reference values of elements in human biological tissues and fluids: a systematic review. Crit Rev Anal Chem 24:363–398.
Cebrian ME, Albores A, Aguilar M (1983) Chronic arsenic poisoning in the north of Mexico. Hum Toxicol 2:121–133.
Chandra Sekhar K, Chary NS, Kamala CT, Venkateswara Rao J, Balaram V, Anjaneyulu Y (2003) Risk assessment and pathway study of arsenic in industrially contaminated sites of Hydrabad: a case study. Environ Int 29:601–611.
Chaudhary K, Ehmann WD, Rengan K, Markesbery WR (1995) Trace elements correlations with age and sex in human fingernails. J Radioanal Nucl Chem Art 195: 51–56.
Chen KL, Amarasiriwardena CJ, Christiani DC (1999) Determination of total arsenic concentrations in nails by inductively coupled plama mass spectrometry. Biol Trace Elem Res 67(2):109–125.
Cheng TP, Morris JS, Koirtyohann SR (1995) Study of the correlation of trace elements in carpenter's toenails. J Radioanal Nucl Chem 195:31–42.
Chiou HY, Hsueh YM, Hsieh LL, Hsu LI, Hsu YH, Hsieh FI (1997) Arsenic methylation capacity, body retention, and null genotypes of glutathione S-transferase MI and TI among current arsenic-exposed residents in Taiwan. Mutat Res 387:197–207.
Czarnowski W, Krechniak J (1990) Fluoride in the urine, hair, and nails of phosphate fertilizer workers. Br J Ind Med 47 (5):349–351.
Daniel CR, Piraccini BM, Tosti A (2004) The nail and hair in forensic science. J Am Acad Dermatol 50(2):258–261.
Das D, Chatterjee A, Mandal BK, Samanta G, Chakraborti D, Chanda B (1995) Arsenic in ground water in six districts of West Bengal, India: the biggest arsenic calamity in the world. Part 2: Arsenic concentration in drinking water, hair, nails, urine, skin-scale and liver tissue (biopsy) of the affected people. Analyst 120:917–924.
Djaldetti M, Fishman P, Hart J (1987) The iron content of fingernails in iron deficient patients. Clin Sci (Lond) 72(6):669–672.
el-Khatib AM, el-Mohandes A (1992) A study of some trace elements in fingernail and hair of Egyptian bilharzial patients using short neutron activation. J Egypt Public Health Assoc 67(3–4):479–490.

Elkins HP (1954) Analyses of biological materials as indices of exposure to organic solvents. AMA Arch Ind Hyg Occup Med 9:212–222.

Escobar H, Arroyo M, Suarez L, Camarero C, Crespo E, Vera C (1980) Copper and zinc levels in nails of children with cystic fibrosis, carriers and healthy controls. An Esp Pediatr 13(2):127–132.

Feldman RG, Niles CA, Kelly-Hayes M, Sax DS, Dixon WJ, Thompson DJ, Landau E (1979) Peripheral neuropathy in arsenic smelter workers. Neurology 29(7): 939–944.

Feskanich D, Owusu W, Hunter DJ, Willett W, Ascherio A, Spiegelman D, Morris S, Spate VL, Colditz G (1998) Use of toenail fluoride levels as an indicator for the risk of hip and forearm fractures in women. Epidemiology 9(4):412–416.

Fite LE, Wainerdi RE, Harrison GM, Doggett GM (1972) Copper analysis of nail clippings for detection of cystic fibrosis. Nuclear activation techniques in life science symposium, IAEA, Vienna, pp 487–499.

Forrai G, Kasperek K, Salamon A, Feinendegen LE (1984) Estimation of zinc and other trace elements in the nails of Hungarian adult twin pairs by neutron activation analysis. Acta Biochim Biophys Acad Sci Hung 19(3–4): 299–304.

Forslind B, Wroblewski R, Afzelius BA (1976) Calcium and sulfur location in human nail. J Invest Dermatol 67(2):273–275.

Gammelgaard B, Veien NK (1990) Nickel in nails, hair and plasma from nickel-hypersensitive women. Acta Derm Venereol 70(5):417–420.

Gammelgaard B, Peters K, Menne T (1991) Reference values for the nickel concentration in human finger nails. J Trace Elem Electrol Health Dis 5(2):121–123.

Garland M, Morris JS, Rosner BA, Stampfer MJ, Spate VL, Baskett CJ, Willett WC, Hunter DJ (1994) Toenail trace element levels as biomarkers: reproducibility over a 6-year period. Cancer Epidemiol Biomark Prev 3(6):523.

Garland M, Morris JS, Stampfer MJ, Colditz GA, Spate VL, Baskett CK, Rosner B, Speizer FE, Willett WC, Hunter DJ (1995) Prospective study of toenail selenium levels and cancer among women. J Natl Cancer Inst 87(7):497–505.

Garland M, Morris JS, Colditz GA, Stampfer MJ, Spate VL, Baskett CK, Rosner B, Speizer FE, Willett WC, Hunter DJ (1996) Toenail trace element levels and breast cancer: a prospective study. Am J Epidemiol 144(7):653–660.

Gerhardsson L, Brune D, Lundstrom NG, Nordberg G, Wester PO (1993) Biological specimen bank for smelter workers. Sci Total Environ 139–140:157–173.

Gerhardsson L, Englyst V, Lundstrom NG, Nordberg G, Sandberg S, Steinvall F (1995) Lead in tissues of deceased lead smelter workers. J Trace Elem Med Biol 9(3): 136–143.

Ghadirian P, Maisonneuve P, Perret C, Kennedy G, Boyle P, Krewski D, Lacroix A (2000) A case-control study of toenail selenium and cancer of the breast, colon, and prostate. Cancer Detect Prev 24(4):305–313.

Gibson RS (1989) Assessment of trace element status in humans. Prog Food Nutr Sci 13(2):67–111.

Guallar E, Sanz-Gallardo MI, van't Veer P, Bode P, Aro A, Gomez-Aracena J, Kark JD, Riemersma RA, Martin-Moreno JM, Kok FJ (2002) Mercury, fish oils, and the risk of myocardial infarction. N Engl J Med 347(22):1747–1754.

Gulson BL (1996) Nails: concern over their use in lead exposure assessment. Sci Total Environ 177:323–327.

Hadjimakos DM, Sheaver TR (1973) Selenium content of human nails: a new index for epidemologic studies of dental caries. J Dent Res 52:389.

Harrington J, Middaugh J, Morse D (1978) A survey of a population exposed to high concentrations of arsenic in well water in Fairbanks, Alaska. Am J Epidemiol 108: 377–384.

Harrison WW, Tyree AB (1971) The determination of trace elements in human fingernails by atomic absorption spectroscopy. Clin Chim Acta 31:63–73.

Hartman TJ, Taylor PR, Alfthan G, Fagerstrom R, Virtamo J, Mark SD, Virtanen M, Barrett MJ, Albanes D (2002) Toenail selenium concentration and lung cancer in male smokers (Finland). Cancer Causes Control 13(10):923–928.

Hayashi M, Yamamoto K, Yoshimura M, Hayashi H, Shitara A (1993) Cadmium, lead, and zinc in human fingernails. Bull Environ Contam Toxicol 50:547–553.

Hemond HF, Solo-Gabriele HM (2004) Children's exposure to arsenic from CCA-treated wooden decks and playground structures. Risk Anal 24(1):51–64.

Henke G, Nucci A, Queiroz LS (1982) Detection of repeated arsenical poisoning by neutron activation analysis of food nail segments. Arch Toxicol 50(2):125–131.

Hewitt DJ, Millner GC, Nye AC, Simmons HF (1995) Investigation of arsenic exposure from soil at a superfund site. Environ Res 68(2):73–81.

Hinwood AL, Sim MR, Jolley D, de Klerk N, Bastone EB, Gerostamoulos J, Drummer OH (2003) Hair and toenail arsenic concentrations of residents living in areas with high environmental arsenic concentrations. Environ Health Perspect 111(2): 187–193.

Hira CK, Partal K, Dhillon KS (2004) Dietary selenium intake by men and women in high and low selenium areas of Punjab. Public Health Nutr 7(1):39–43.

Hopps HC (1977) The biologic bases for using hair and nails for analyses of trace elements. Sci Total Environ 7(1):71–89.

Horn-Ross PL, Morris JS, Lee M, West DW, Whittemore AS, McDougall IR, Noweis K, Stewart SL, Spate VL, Shiau AC, Krone MR (2001) Iodine and thyroid cancer risk among women in a multiethnic population: the Bay Area Thyroid Cancer Study. Cancer Epidemiol Biomark Prev 10(9):979–985.

Hunter DJ, Morris JS, Chute CG, Kushner E, Colditz GA, Stampfer MJ, Speizer FE, Willett WC (1990a) Predictors of selenium concentration in human toenails. Am J Epidemiol 132(1):114–122.

Hunter DJ, Morris JS, Stampfer MJ, Colditz GA, Speizer FE, Willett WC (1990b) A prospective study of selenium status and breast cancer risk. JAMA 264(9):1128–1131.

Ilhan A, Ozerol E, Gulec M, Isik B, Ilhan N, Ilhan N, Akyol O (2004) The comparison of nail and serum trace elements in patients with epilepsy and healthy subjects. Prog Neuropsychopharmacol Biol Psychiatry 28(1):99–104.

Iyengar GV, Woittiez J (1988) Trace elements in human clinical specimen: evaluation of literature data to identify reference values. Clin Chem 34(3):474–481.

Kanabrocki EL, Kanabrocki JA, Greco J, Kaplan E, Oester YT, Brar SS, Gustafson PS, Nelson DM, Moore CE (1979) Instrumental analysis of trace elements in thumbnails of human subjects. Sci Total Environ 13(2):131–140.

Kao HF (1990) Quantitation of calcium levels in the nails of psoriasis patients by energy dispersive X-ray microanalysis. J Formosa Med Assoc 89(5):363–365.

Kapito L, Shwachman H (1964) Spectroscopic analysis of tissues from patients with cystic fibrosis and controls. Nature (Land) 202:501–502.

Kapito L, Makdoodian RRW, Townley W, Khaw KJ, Shwachman H (1965) Studies of cystic fibrosis: analysis of nail clippings for sodium and potassium. N Engl J Med 272:504–509.

Karagas MR, Morris JS, Weiss JE, Spate V, Baskett C, Greenberg ER (1996) Toenail samples as an indicator of drinking water arsenic exposure. Cancer Epidemiol Biomark Prev 5(10):849–852.

Karagas MR, Tosteson TD, Blum J, Klaue B, Weiss JE, Stannard V, Spate V, Morris JS (2000) Measurement of low levels of arsenic exposure: a comparison of water and toenail concentrations. Am J Epidemiol 152(1):84–90.

Karagas MR, Le CX, Morris S, Blum J, Lu X, Spate V, Carey M, Stannard V, Klaue B, Tosteson TD (2001) Markers of low level arsenic exposure for evaluating human cancer risks in a US population. Int J Occup Med Environ Health 14(2):171–175.

Kardinaal AF, Kok FJ, Kohimeier L, Martin-Moreno JM, Ringstad J, Gomez-Aracena J, Mazaev VP, Thamm M, Martin BC, Aro A, Kark JD, Delgado-Rodriguez M, Riemersma RA, van't Veer P, Huttunen JK (1997) Association between toenail selenium and risk of acute myocardial infaction in European men. The EURAMIC Study. European Antioxidant Myocardial Infarction and Breast Cancer. Am J Epidemiol 145(4):373–937.

Karita K, Takano T (1994) Relation of fingernail mineral concentrations to bone mineral density in elderly Japanese women. Nippon Koshu Eisei Zasshi 41(8):759–763.

Karita K, Takano T, Nakamura S, Haga N, Iwaya T (2001) A search for calcium, magnesium and zinc levels in fingernails of 135 patients with osteogenesis imperfecta. J Trace Elem Med Biol 15(1):36–39.

Karpas Z (2001) Uranium bioassay-beyond urinalysis. Health Phys 81(4):460–463.

Kasperek K, Iyengar GV, Feinendegen LE, Hashish S, Mahfouz M (1982) Multielement analysis of fingernail, scalp hair and water samples from Egypt (a preliminary study). Sci Total Environ 22(2):149–168.

Katayama Y, Ishida N (1987) Determination of antimony in nail and hair by thermal neutron activation analysis. Radioisotopes 36(3):103–107.

Kok FJ, Hofman A, Witteman JC, de Bruijn AM, Kruyssen DH, de Bruin M, Valkenburg HA (1989) Decreased selenium levels in acute myocardial infarction. JAMA 261(8): 1161–1164.

Krogh V, Pala V, Vinceti M, Berrino F, Ganzi A, Micheli A, Muti P, Vescovi L, Ferrari A, Fortini K, Sieri S, Vivoli G (2003) Toenail selenium as biomarker: reproducibility over a one-year period and factors influencing reproducibility. J Trace Elem Med Biol (Suppl) 1:31–36.

Kucera J, Bencko V, Papayova A, Saligova D, Tejral J, Borska L (2001) Monitoring of occupational exposure in manufacturing of stainless steel constructions. Part I: Chromium, iron, manganese, molybdenum, nickel and vanadium in the workplace air of stainless steel welders. Cent Eur J Public Health 9(4):171–175.

Lapatto R, Hietamaki A, Raisanen J (1989) Quantitative trace element analysis of human nails with external beam PIXE. Biol Trace Elem Res 19(3):161–170.

Leon-Espinosa de los Monteros MT, Gil Extremera B, Maldonado Martin A, Luna del Castillo JD, Munox Parra F, Ruiz Lopez MF, Huertas Hernandez F, Cobo Martinez F (2000) Zinc and chronic obstructive pulmonary disease. Rev Clin Esp 200(12):649–653.

Lim P, Tay SB, Tan IK (1972) Nail calcium and magnesium in chronic uremia. Clin Chim Acta 42:47–49.

Lin TH, Huang YL, Wang MY (1978) Arsenic species in drinking water, hair, fingernails, and urine of patients with black foot disease. J Toxicol Environ Health A 53(2):85–93.

Longnecker MP, Taylor PR, Levander OA, Howe M, Veillon C, McAdam PA, Patterson KY, Holden JM, Stampfer MJ, Morris JS (1991) Selenium in diet, blood, and toenails in relation to human health in a seleniferous area. Am J Clin Nutr 53(5):1288–1294.

Longnecker MP, Stampfer MJ, Morris JS, Spate V, Baskett C, Mason M, Willett WC (1993) A 1-y trial of the effect of high-selenium bread on selenium concentrations in blood and toenails. Am J Clin Nutr 57(3):408–413.

Longnecker MP, Stram DO, Taylor PR, Levander OA, Howe M, Veillon C, McAdam PA, Patterson KY, Holden JM, Morris JS, Swanson CA, Willett WC (1996) Use of selenium concentration in whole blood, serum, toenails, or urine as a surrogate measure of selenium intake. Epidemiology 7(4):384-390.

Lubach D, Wurzinger R (1986) Trace elements in samples of brittle and nonbrittle fingernails. Derm Beruf Umwelt 34(2):37–39.

M'Baku SB, Parr RM (1982) Interlaboratory study of trace and other elements in the IAEA powdered human hair reference materials, H.H.I. J Radioanal Chem 69(1): 171–180.

MacIntosh DL, Williams PL, Hunter DJ, Sampson LA, Morris SC, Willett WC, Rimm EB (1997) Evaluation of a food frequency questionnaire-food composition approach for estimating dietary intake of inorganic arsenic and methylmercury. Cancer Epidemiol Biomark Prev 6(12):1043–1050.

Madzhunov N, Zaprianov Z, Georgieva R (1993) Skin reactivity and the biological assessment of chromium exposure. Probl Khig 18:167–175.

Mahler DJ, Scott AF, Walsh JR, Haynie G (1970) A study of trace metals in fingernails and hair using neutron activation analysis. J Nucl Med 11:739–742.

Majumdar S, LeonChatterjee J, Chaudhuri K (1999) Ultra structural and trace metal studies on radiographers' hair and nails. Biol Trace Elem Res 67(2):127–138.

Mandal BK, Ogra Y, Suzuki KT (2003) Speciation of arsenic in human nail and hair from arsenic-affected area by HPLC-inductively coupled argon plasma mass spectrometry. Toxicol Appl Pharmacol 189(2):73–83.

Mannisto S, Alfthan G, Virtanen M, Kataja V, Uusitupa M, Pietinen P (2000) Toenail selenium and breast cancer: a case-control study in Finland. Eur J Clin Nutr 54(2): 98–103.

Martin GM (1964) Copper content of hair and nails of normal individuals and patients with hepatolenticular degeneration. Nature (Lond) 202:903–904.

Martin-Moreno JM, Gorgojo L, Riemersma RA, Gomez-Aracena J, Kark JD, Guillen J, Jimenez J, Ringstad JJ, Fernandez-Crehuet J, Bode P, Kok FJ (2003) Myocardial infarction risk in relation to zinc concentration in toenails. Br J Nutr 89(5):673–678.

Marumo F, Tsukamoto Y, Iwanami S, Kishimoto T, Yamagami S (1984) Trace element concentrations in hair, fingernails and plasma of patients with chronic renal failure on hemodialysis and hemofiltration. Nephron 38(4):267–272.

Masironi R, Koirtyohann SR, Pierce JO, Schamschula RG (1976) Calcium content of river water, trace element concentrations in toenails, and blood pressure in village populations in New Guinea. Sci Total Environ 6(1):41–53.

McCurdy RF, Hindmarsh JT (1987) Determination of arsenic and its derivatives in biological samples. Proceedings of the Seminar on Clinical and Analytical Toxicology, Association of Clinical Sciences, pp 95–103.

McKenzie JM (1979) Content of zinc in serum, urine, hair, and toenails of New Zealand adults. Am J Clin Nutr 32(3):570–579.

McKenzie JM, Guthrie BE, Prior IA (1978) Zinc and copper status of Polynesian residents in the Tokelau Islands. Am J Clin Nutr 31(3):422–428.

Morris SJ, Stampfer MJ, Willett W (1983) Dietary selenium in humans: toenails as an indicator. Biol Trace Elem Res 5:529–537.

Mortada WI, Sobh MA, El-Dfrawy MM, Farahat SE (2001) Study of lead exposure from automobile exhaust as a risk for nephrotoxicity among traffic policemen. Am J Nephrol 21(4):274–279.

Mortada WI, Sobh MA, el-Defrawy MM, Farahat SE (2002) Reference intervals of cadmium, lead, and mercury in blood, urine, hair, and nails among residents in Mansoura City, Nile delta, Egypt. Environ Res 90(2):104–110.

Mortada WI, Sobh MA, El-Defrawy MM (2004) The exposure to cadmium, lead and mercury from smoking and its impact on renal integrity. Med Sci Monit 10(3):CR112–CR116.

Narang AP, Chawla LS, Khurana SB (1987) Levels of arsenic in Indian opium eaters. Drug Alcohol Depend 20(2):149–153.

Ndiokwere C (1985) A survey of arsenic levels in human hair and nails: exposure of wood treatment factory employees in Nigeria. Environ Pollut 9:95–105.

Nichols TA, Morris JS, Mason MM (1998) The study of human nails as an intake monitor for arsenic using neutron activation analysis. J Radioanal Chem 236:51–56.

Nielsen NH, Kristiansen J, Borg L, Christensen JM, Poulsen LK, Menne T (2000) Repeated exposures to cobalt or chromate on the hands of patients with hand eczema and contact allergy to that metal. Contact Dermatitis 43(4):212–215.

Nowak B, Chmielnicka J (2000) Relationship of lead and cadmium to essential elements in hair, teeth, and nails of environmentally exposed people. Ecotoxicol Environ Saf 46(3): 265–274.

Olguin A, Jauge P, Cebrian M, Albores A (1983) Arsenic levels in blood, urine, hair and nails from a chronically exposed human population. Proc West Pharmacol Soc 26:175–177.

Oluwole AF, Asubiojo OI, Adekile AD, Filby RH, Bragg A, Grimm CI (1990) Trace element distribution in the hair of some sickle cell anemia patients and controls. Biol Trace Elem Res 26–27:479–484.

Oluwole AF, Ojo JO, Durosinmi MA, Asubiojo OI, Akanle OA, Spyrou NM, Filby RH (1994) Elemental composition of head hair and fingernails of some Nigerian subjects. Biol Trace Elem Res 43–45:443–452.

Othman I, Spyrou NM (1980) The abundance of some elements in hair and nail from the Machakos District of Kenya. Sci Total Environ 16(3):267–278.

Ovaskainen ML, Virtamo J, Alfthan G, Haukka J, Pietinen P, Taylor PR, Huttunen JK (1993) Toenail selenium as an indicator of selenium intake among middle-aged men in an area with low soil selenium. Am J Clin Nutr 57(5):662–665.

Pallotti G, Bencivenga B, Simonetti T (1979) Total mercury levels in whole blood, hair and fingernails for a population group from Rome and its surroundings. Sci Total Environ 11(1):69–72.

Parr RM, Schelenz R, Bellestra S (1988) IAEA Biological Reference Materials. Proceedings of the Third International Symposium on Biological Reference Materials. Bayreuch FRG.

Peters K, Gammelgaard B, Menne T (1991) Nickel concentrations in fingernails as a measure of occupational exposure to nickel. Contact Dermatitis 25(4):237–241.

Platz EA, Helzlsouer KJ, Hoffman SC, Morris JS, Baskett CK, Comstock GW (2002) Prediagnostic toenail cadmium and zinc and subsequent prostate cancer risk. Prostate 1:52(4):288–296.

Rakovi M, Foltynova V, Pilecka N, Glagolicova A, Kucera J (1997) Assessment of metals and metalloids in skin derivatives of volunteers from capital city of Prague, Czech Republic. Sb Lek 98(2):107–114.

Rayman MP, Bode P, Redman CW (2003) Low selenium status is associated with the occurrence of the pregnancy disease preeclampsia in women from the United Kingdom. Am J Obstet Gynecol 189(5):1343–1349.

Ritchie KA, Gilmour WH, Macdonald EB, Burke FJ, McGowan DA, Dale IM, Hammersley RM, Binnie V, Collington D (2002) Health and neuropychological functioning of dentists exposed to mercury. Occup Environ Med 59(2):287–293.

Robson JRK, Brooks GJ (1974) The distribution of calcium in finger nails from healthy and malnourished chidren. Clin Chim Acta 55:255–257.

Rodushkin I, Axelsson MD (2000a) Application of double focusing sector field ICP-MS for multi-elemental characterization of human hair and nails. Part I. Analytical methodology. Sci Total Environ 250:83–100.

Rodushkin I, Axelsson MD (2000b) Application of double focusing sector field ICP-MS for multi-elemental characterization of human hair and nails. Part II. A study of the inhabitants of northern Sweden. Sci Total Environ 343(24):21–36.

Rogers MA, Thomas DB, Davis S, Weiss NS, Vaughan TL, Nevissi AE (1991) A case-control study of oral cancer and pre-diagnostic concentrations of selenium and zinc in nail tissue. Int J Cancer 48(2):182–188.

Rogers MA, Thomas DB, Davis S, Vaughan TL, Nevissi AE (1993) A case-control study of element levels and cancer of the upper aero-digestive tract. Cancer Epidemiol Biomark Prev 2(4):305–312.

Roomans GM, Afzelius BA, Kollberg H, Forslind B (1978) Electrolytes in nails analysed by X-ray microanalysis in electron microscopy. Considerations on a new method for the diagnosis of cystic fibrosis. Acta Paediatr Scand 67(1):89–94.

Sabbioni E, Minoia C, Pietra R, Mosroni G, Forni A, Scansetti G (1994) Metal determinations in biological specimens of diseased and non-diseased hard metal workers. Sci Total Environ 150(1–3):41–54.

Saha A, Sadhu HG, Karnik AB, Patel TS, Sinha SN, Saiyed HN (2003) Erosion of nails following thallium poisoning: a case report. Occup Environ Med 61(7): 640–642.

Samanta G, Sharma R, Roychowdhury T, Chakraborti D (2004) Arsenic and other elements in hair, nails, and skin-scales of arsenic victims in West Bengal, India. Sci Total Environ 326(1):33–47.

Sato S (1991) Iron deficiency: structural and microchemical changes in hair, nails, and skin. Semin Dermatol 10(4):313–319.

Schamschula RG, Duppenthaler JL, Sugar E, Un PS, Toth K, Barmes DE (1988) Fluoride intake and utilization by Hungarian children: associations and interrelationships. Acta Physiol Hung 72(2):253–261.

Shrestha RR, Shrestha MP, Upadhyay NP, Pradhan R, Khadha R, Maskey A, Maharjan M, Tuladhar S, Dahal BM, Shrestha K (2003) Groundwater arsenic contamination, its health impact and mitigation program in Nepal. J Environ Sci Health Part A Toxicol Hazard Subst Environ Eng 38(1):185–200.

Sobolewski S, Lawrence AC, Bagshaw P (1978) Human nails and body iron. J Clin Pathol 31(11):1068–1072.

Sohler A, Wolcott P, Pfeiffer CC (1976) Determination of zinc in fingernails by non-flame atomic absorption spectroscopy. Clin Chim Acta 70(3):391–398.

Spate VL, Morris JS, Chickos S, Baskett CK, Mason MM, Cheng TP, Reams CL, Furnee C, Willett W, Horn-Ross P (1995) Determination of iodine in human nails via epithermal neutron activation analysis. J Radioanal Nucl Chem 195:21–30.

Subramanian R, Sukumar A (1988) Biological reference materials and analysis of toxic elements. Fresenius Z Anal Chem 332:623–626.
Sukumar A (2002) Factors influencing levels of trace elements in human hair. Rev Environ Contam Toxicol 175:47–78.
Sukumar A, Subramanian R (1992a) Elements in hair and nails of residents from a village adjacent to New Delhi: influence of place of occupation and smoking habits. Biol Trace Elem Res 34:99–105.
Sukumar A, Subramanian R (1992b) Elements in hair and nails of urban residents of New Delhi: CHD, hypertensive and diabetic cases. Biol Trace Elem Res 34:89–98.
Suzuki T, Watanabe S, Matsuo N (1989) Comparison of hair with nail as index media for biological monitoring of mercury. Sangyo Igaku 31(4):235–238.
Swanson CA, Longnecker MP, Veillon C, Howe M, Levander OA, Taylor PR, McAdam PA, Brown CC, Stampfer MJ, Willett WC (1990) Selenium intake, age, gender, and smoking in relation to indices of selenium status of adults residing in a seleniferous area. Am J Clin Nutr 52(5):858–862.
Takagi Y, Matsuda S, Imai S, Ohmori Y, Masuda T, Vinson JA, Mehra MC, Puri BK, Kaniewski A (1988) Survey of trace elements in human nails: an international comparison. Bull Environ Contam Toxicol 41(5):690–695.
Tang YR, Zhang SQ, Xiong Y, Zhao Y, Fu H, Zhang HP, Xiong KM (2003) Studies of five microelement contents in human serum, hair, and fingernails correlated with aged hypertension and coronary heart disease. Biol Trace Elem Res 92(2):97–104.
Tsolova S, Zaprianov Z, Dmitrov TS, Georgieva R, Khinkova L, Petrov I, Nikolova L (1995) The assessment of arsenic exposure in copper smelting. Probl Khig 20:128–138.
van den Brandt PA, Goldbohm RA, van't Veer P, Bode P, Dorant E, Hermus RJ, Sturmans F (1994) Toenail selenium levels and the risk of breast cancer. Am J Epidemiol 140(1):20–26.
van den Brandt PA, Goldbohm RA, van't Veer P, Bode P, Dorant E, Hermus RJ, Sturmans F (1993a) A prospective cohort study on selenium status and the risk of lung cancer. Cancer Res 53(20):4860–4865.
van den Brandt PA, Goldbohm RA, van't Veer P, Bode P, Dorant E, Hermus RJ, Sturmans F (1993b) A prospective cohort study on toenail selenium levels and risk of gastrointestinal cancer. J Natl Cancer Inst 85(3):224–229.
van den Brandt PA, Zeegers MP, Bode P, Goldbohm RA (2003) Toenail selenium levels and the subsequent risk of prostate cancer: a prospective cohort study. Cancer Epidemiol Biomark Prev 12(9):866–871.
van Noord PA, Collette HJ, Maas MJ, de Waard F (1987) Selenium levels in nails of premenopausal breast cancer patients assessed prediagnostically in a cohort-nested case-referent study among women screened in the DOM project. Int J Epidemiol 16(2):318–322.
van Stekelenburg GJ, van de Laar AJ, van der Laag J (1975) Copper analysis of nail clippings. An attempt to differentiate between normal children and patients suffering from cystic fibrosis. Clin Chim Acta 59(2):233–240.
van't Veer P, van der Wielen RP, Kok FJ, Hermus RJ, Sturmans F (1990) Selenium in diet, blood, and toenails in relation to breast cancer: a case-control study. Am J Epidemiol 131(6):987–994.
Vance DE, Ehmann WD, Markesbery WR (1988) Trace element content in fingernails and hair of a non-industrialized US control population. Biol Trace Elem Res 17:109–121.
Vance DE, Ehmann WD, Markesbery WR (1990) A search for longitudinal variations in trace element levels in nails of Alzheimer's disease patients. Biol Trace Elem Res 26–27:461–470.

Vannucchi H, Favaro RMD, Cunha DF, Marchini JS (1995) Assessment of zinc nutritional status of pellagra patients. Alcohol Alcoholism 30(3):297–302.

Vecht-Hart CM, Bode P, Trouerbach WT, Collette HJ (1995) Calcium and magnesium in human toenails do not reflect bone mineral density. Clin Chim Act 236(1):1–6.

Waritz RS (1979) Biological indicator of chemical dosage and burden. In: Cralley IV, Cralley LJ (eds) Patty's Industrial Hygiene and Toxicology, vol III. Wiley, New York, pp 257–318.

Weismann K, Hoyer H (1982) Zinc deficiency dermatoses. Etiology, clinical aspects and treatment. Hautarzt 33(8):405–410.

Wilhelm M, Hafner D, Lombeck I, Ohnesorge FK (1991) Monitoring of cadmium, copper, lead and zinc status in young children using toenails: comparison with scalp hair. Sci Total Environ 103(2–3):199–207.

Wilhelm M, Lombeck I, Ohnesorge FK (1994) Cadmium, copper, lead and zinc concentrations in hair and toenails of young children and family members: a follow-up study. Sci Total Environ 141(1–3):275–280.

Yoshizawa K, Willett WC, Morris SJ, Stampfer MJ, Spiegelman D, Rimm EB, Giovannucci E (1998) selenium level in toenails and the risk of advanced prostate cancer. J Natl Cancer Inst 90(16):1219–1224.

Yoshizawa K, Rimm EB, Morris JS, Spate VL, Hsieh CC, Spiegelman D, Stampfer MJ, Willett WC (2002) Mercury and the risk of coronary heart disease in men. N Engl J Med 347(22):1755–1760.

Yoshizawa K, Ascherio A, Morris JS, Stampfer MJ, Giovannucci E, Baskett CK, Willett WC, Rimm EB (2003) Prospective study of selenium levels in toenails and risk of coronary heart disease in men. Am J Epidemiol 158(9):852–860.

Zeegers MP, Goldbohm RA, Bode P, van den Brandt PA (2002) Prediagnostic toenail selenium and risk of bladder cancer. Cancer Epidemiol Biomark Prev 11(11):1292–1297.

Zevin D, Weinstein T, Levi J, Djaldetti M (1991) X-ray microanalysis of the fingernails of uremic patients treated by hemodialysis. Clin Nephrol 36(6):302–304.

Zhang N, Tian S, Zhang S (2001) Determination of trace cadmium in human hair, fingernail and toenail for ten years by flame atomic absorption spectrometry. Guang Pu Xue Yu Guang Pu Fen Xi 21(3):393–396.

Manuscript received December 23, 2004; accepted January 26, 2005.

Index

MIX
Papier aus verantwortungsvollen Quellen
Paper from responsible sources
FSC® C105338

If you have any concerns about our products,
you can contact us on
ProductSafety@springernature.com

In case Publisher is established outside the EU,
the EU authorized representative is:
Springer Nature Customer Service Center GmbH
Europaplatz 3, 69115 Heidelberg, Germany

Printed by Libri Plureos GmbH
in Hamburg, Germany